Grover Coe

**Concentrated Organic Medicines**

Being a practical exposition of the therapeutic properties and clinical employment

of the combined proximate medicinal constituents of indigenous and foreign

plants, to which is added a brief history of crude organic

Grover Coe

**Concentrated Organic Medicines**
*Being a practical exposition of the therapeutic properties and clinical employment of the combined proximate medicinal constituents of indigenous and foreign plants, to which is added a brief history of crude organic*

ISBN/EAN: 9783337314491

Printed in Europe, USA, Canada, Australia, Japan

Cover: Foto ©berggeist007 / pixelio.de

More available books at **www.hansebooks.com**

# CONCENTRATED ORGANIC MEDICINES:

BEING A

PRACTICAL EXPOSITION

OF THE

## THERAPEUTIC PROPERTIES

AND

## CLINICAL EMPLOYMENT

OF THE COMBINED PROXIMATE MEDICINAL CONSTITUENTS OF

## INDIGENOUS AND FOREIGN PLANTS.

TO WHICH IS ADDED A BRIEF HISTORY OF CRUDE ORGANIC REMEDIES, CONSTITUENTS OF PLANTS, CONCENTRATED MEDICINES, OFFICINAL PREPARATIONS, &C., &C.

BY GROVER COE M. D.

NEW YORK:
PUBLISHED BY B. KEITH & CO., 41 LIBERTY STREET.

# CONTENTS.

## PART I.

### CHAPTER I.

#### CRUDE ORGANIC REMEDIES.

Objections to their use—Of uncertain value—Frequently inert—Facts of analysis—Influences of soil, climate, cultivation, kiln-drying, etc.

### CHAPTER II.

#### CONSTITUENTS OF PLANTS.

Acids—Alkaloids—Indifferent or neutral substances—Cellulose—Xylogen—Cuticular or cork substance—Protein—Amylum—Dextrine—Sugars—Pectin—Gum—Mucilage—Muciresins—Viscin—Inorganic elements—Fixed oils—Wax—Volatile oils—Camphors—Resins—Oleo-Resins—Gum-Resins—Resinoids—Caoutchouc—Coloring matters—Extractive substances or neutrals—Humus—Apotheme—Fermentation—Decomposition—Putrefaction—Amygdalin—Emulsin—etc., etc.

### CHAPTER III.

#### CONCENTRATED MEDICINES.

Officinal preparations—Infusions—Decoctions—Extracts—Aqueous, Alcoholic, Hydro-alcoholic, Inspissated, Fluid, etc.—their liability to decomposition—of variable strength—frequently inert, etc.—Concentrated Medicines Proper—their advantage—uniform and definite in strength—not liable to change—methods of administration, etc., etc.

## PART II.

#### CONCENTRATED MEDICINES PROPER.

Their therapeutic and clinical history—Senecin—Asclepin—Gelsemin—Macrotin—Ampelopsin—Geranin—Populin—Cypripedin—Chimaphilin—Dioscorein—Chelonin—Helonin—Leptandrin—Digitalin—Rhusin—Baptisin—Podophyllin—Myricin—Euonymin—Alnuin—Viburnin—Cornin—Rumin—Caulophyllin—Jalapin—Phytolacin—Hyoscyamin—Stillingia—Lupulin—Veratrin—Eupatorin Perfo.—Eupatoria Purpu.—Corydalin—Juglandin—Trilliin—Scutellarin—Apocynin—Irisin—Hydrastin—Hamamelin—Euphorbin—Lycopin—Fraserin—Xanthoxylin—Aconitin—Colocynthin—Rhein—Atropin—Boptisin—etc., etc.

# TABLE OF AVERAGE DOSES.

| Powders. | grs. |
|---|---|
| Alnuin | ½ to 10 |
| Ampelopsin | 2 to 5 |
| Apocynin | ½ to 3 |
| Asclepin | 1 to 5 |
| Baptisin | 1 to 3 |
| Caulophyllin | 2 to 5 |
| Cerasein | 2 to 10 |
| Chelonin | 2 to 5 |
| Chimaphilin | 2 to 5 |
| Collinsonin | 2 to 5 |
| Cornin | 2 to 5 |
| Corydalin | 1 to 3 |
| Cypripedin | 2 to 4 |
| Digitalin | ⅛ to ½ |
| Dioscorein | 2 to 5 |
| Euonymin | 1 to 4 |
| Euphorbin | 1 to 3 |
| Eupatorin Perfo. | 1 to 4 |
| Eupatorin Purpu. | 2 to 5 |
| Fraserin | 2 to 10 |
| Gelsemin | ½ to 2 |
| Geranin | 2 to 5 |
| Hamamelin | 1 to 3 |
| Helonin | 2 to 5 |
| Hydrastin | 1 to 3 |
| Hyoscyamin | ⅛ to ½ |
| Irisin | 1 to 3 |
| Jalapin | 2 to 5 |
| Juglandin | 2 to 10 |
| Leptandrin | 2 to 5 |
| Lupulin | 1 to 4 |
| Lycopin | 1 to 4 |
| Macrotin | ½ to 2 |
| Menispermin | 1 to 5 |
| Myricin | 2 to 5 |
| Podophyllin | ½ to 3 |
| Populin | 2 to 5 |
| Prunin | 1 to 3 |
| Phytolacin | ½ to 3 |
| Rhusin | 1 to 3 |
| Rumin | 1 to 3 |

| Powders. | grs. |
|---|---|
| Sanguinarin | ½ to 2 |
| Senecin | 2 to 5 |
| Scutellarin | 2 to 5 |
| Stillingin | 2 to 5 |
| Smilacin | 2 to 5 |
| Strychnin | 1/32 to 1/8 |
| Trilliin | 2 to 5 |
| Veratrin | ½ to 1 |
| Viburnin | 2 to 5 |
| Xanthoxylin | 2 to 5 |

### CONCENTRATED TINCTURES.

| | gtt. |
|---|---|
| Apocynum | 2 to 10 |
| Collinsonia | 5 to 20 |
| Digitalis | 1 to 4 |
| Euonymus | 2 to 10 |
| Eupatorium Purpu. | 2 to 10 |
| Gelseminum | 5 to 20 |
| Hyoscyamus | 5 to 20 |
| Rhus Glab. | 2 to 10 |
| Scutellaria | 2 to 10 |
| Smilax | 10 to 20 |
| Senecio | 2 to 10 |
| Veratrum | 1 to 8 |
| Xanthoxylum | 2 to 5 |

### OILS.

| | |
|---|---|
| Capsicum | ½ to 2 |
| Erigeron | 2 to 10 |
| Lobelia | ½ to 2 |
| Solidago | 2 to 5 |
| Stillingia | ½ to 2 |
| Xanthoxylum | 2 to 5 |
| Con. Comp. Stillingia Alterative | 2 to 10 |
| Wine Tinct. Lobelia:— | |
| As an expectorant | 2 to 10 |
| As an emetic | ℥ ii. ℥ ii. |

# PREFACE.

He is said to be a bold author who writes a preface. Nevertheless, undaunted by the fates of our predecessors, we voluntarily submit to the allegation of hardihood, and bow our head submissively in deference to the omniscient wisdom of the critics.

We remember to have somewhere read a good story of an artist, who, after painting a beautiful picture, bestowing upon it much time and labor, exposed it for criticism in the market place. Beside the picture he placed brushes and a pallet of colors, with a request that all good judges of the art of painting would remedy those defects they might discover which had escaped his own eye.

Nothing could be more gratifying to the innate vanity of such as considered themselves capable of deciding upon the merits or demerits of all they saw, than this general invitation. Every one who looked upon the canvass discovered something essentially wrong in the composition, which was retouched, according to his individual idea of the sublime or the beautiful. The brush was no sooner laid down than another took it up; it was, therefore constantly applied; but when the author called at evening to examine and admire the friendship which had been manifested for his reputation as an artist, not a single vestige of the original design remained. Although all who chose had contributed the pigment they considered absolutely indispensible to perfect the picture, the next day it was unanimously declared that the painter was a man of no ingenuity or knowledge of his profession.

In the arrangement of the little volume now submitted to the profession, the writer has consulted his own notions of propriety, striving to keep in view, at the same time, the best interests of his readers. That it has its imperfections, will be apparent to all who do him the honor to peruse it. Yet we can not emulate the generosity of the artist by supplying the materials with which to remedy apparent defects, although conscious that had the original materials been placed in abler and more experienced hands, a more perfect work would have been secured. Like the artist, we may extend a general invitation to the profession to modify the peculiarities of detail, not for universal application, but for the purpose of meeting the requirements of individual circumstances and necessities.

We have not sought to charm the sense by elegance of diction, nor aimed to delude the reason by ingeniously wrought hypotheses; but simply to present, in a concise manner, an array of scientific facts which we hope will be of practical utility to the profession. We invite attention to the subject matter of our treatise, rather than to a critical examination of the language in which it is embodied. In treating of crude organic remedies, constituents of plants, and officinal preparations, we have presented many new, and, it may be, startling facts; but they are none the less worthy, for this reason, of the serious and impartial consideration of the profession. Therapeutical and pharmaceutical science are at the very foundation of practical medicine, and he who perfects their principles will be the acknowledged benefactor of his race. The field of organic chemical science has been, as yet, but little tilled, and much ground still remains wherein progressive explorations and manipulative skill have inexhaustible resources yet to unfold. But notwithstanding the existence of hidden mines of therapeutic wealth yet undeveloped, much has already been accomplished in bringing forth from the secret recesses of nature's store-house the means wherewith to practice the art of healing. Organic chemistry has solved the problem of vegetable organisms, defined the characteristics that distinguish the *physical* from the *therapeutic*, isolated the motor-excitant constituents from their non-medicinal investure, and furnished the physician with the means of cure, defined in sensible properties, of uniform therapeutic power, and of specific value in fulfilling the indications of disease. The isolation and re-combination of the active constituents of medicinal plants is one of the most important features of modern pharmacy. Instead of isolating a single principle and rejecting the remainder, thus doing violence to the therapeutic integrity of the plant, the aggregate medicmal constituents are now combined in one preparation, and thus we have a pharmaceutical compound of nature's own preparing. Hereby are secured the various therapeutic powers of which the plant, from a consideration of its physiological effects, is known to be possessed.

The reader will perceive, in the second part of this volume, that we have ignored the medical casuistry of those authors who have attempted to disprove, with specious arguments, the existence of certain classes of therapeutic powers. We have faithfully recorded our carefully made observations of the physiological effects of the remedies, and we hold that, when a medicine produces a specific physiological effect, it should be accredited with the power known to be requisite to produce it. It is not for us to determine whether this result be primary or secondary, so long as we can rely upon the certainty of the remedy in fulfilling the indications for which we exhibit it. As Podophyllin promotes the catamenial flow, we award it the possession of an emmenagogue power. As Veratrin lessens the force and frequency of the pulse, when abnormally excited, we term it an arterial sedative, although the question might arise whether the sedative influence is the result of primary or reflex action. In a practical point of view it is sufficient for us to know that it may be relied upon for fulfilling this indication. Gelsemin, Viburnin, Dioscorein and Lobelia relax spasm and control the action of the muscular system, hence we term them, and we think with propriety, anti-spasmodics. But as we have to deal with the practical rather than the theoretical, we leave the solution of the *modus operandi* of medicines to those whose *forte* lies in framing plausible hypotheses.

In the second chapter, in giving the chemical formulas of the various constituents, it will be observed that we have followed the earlier method of single atoms Most of the facts there adduced in relation to vegetable constituents have been elucidated by recent analyses. The history of the resinoid, neutral, and muciresin principles is entirely original, never before having appeared in print, and is the fruit of personal research. We have to acknowledge but little indebtedness to other authors. In some few instances we have consulted Berzelius, Rhind, and other vegetable physiologists and chemists, when controversial points arose, preferring, however, to rely upon recent personal analyses and experiments.

To Adolph Behr, A.M., the gentlemanly and accomplished chemist attached to the laboratory of B. Keith & Co., we are under especial obligations. He has kindly afforded us access to valuable private notes, and materially facilitated our labors by timely furnishing important facts and suggestions. The profession are deeply indebted to the labors of this gentleman for having so successfully elevated the standard of therapeutical and pharmaceutical science.

Our thanks are likewise due to the enterprising publishers, Messrs. B. Keith & Co., for the liberal manner in which they have gotten up this volume, a compliment which, together with the approbation of the profession, rewards us amply for the labor bestowed.

Some few typographical errors have undoubtedly crept in, consequent upon family afflictions and professional cares, by which our attention has been much diverted, but we trust to the kindly consideration of our readers to overlook the mechanical defects. The text we have carefully revised, and believe it to be free from any serious imperfections.

And now we commit our little volume to the hands of a conservative, yet liberal profession, confident that they will impartially consider the substance of our exposition, and neither approve nor condemn except in accordance with the rigid requirements of experimental science.

If we have herein recorded a single additional truth that shall be deemed worthy of being employed in rearing the superstructure of medical science; if we have suggested aught that shall enable our professional brethren to smooth a single wrinkle in the pillow of care, or check the coursing of a single tear down the furrowed cheek of suffering humanity, we will accept the token, with gratitude, as the full measure of our reward.  G. C.

New York, Sept., 1858.

# PREFACE TO THE SECOND EDITION.

The urgent demands of the profession rendering a second edition of the present volume necessary, the author avails himself of the opportunity to express his sincere acknowledgments for the favor with which his feeble attempt to elevate the standard of Materia Medica has been received. Progression is the order of the day, and in no department of medical science is its operations more manifest than in that of therapeutics. The writer hopes to see, at no distant day, the formation of an indigenous Materia Medica, competent for the wants of all, and at once the pride and glory of our common country.

To the present volume has been added the history of several agents not unknown to the profession, but for the first time presented in their present form. The characteristic difference is the same as that of the other concentrated preparations described in this work, namely, the isolation and recombination of the several active constituents resident in each plant. Their clinical history has been drawn from competent sources, upon which the writer has been compelled to rely in the absence of satisfactory personal observation. The authority, however, is as much entitled to credence as would be the vouching of the author's own experience.

This addition has been made in the form of addenda, with a view to an ultimate revision of the entire work together with the rendition of whatever valuable original information upon the subject of organic remedies may have been at such time developed. The writer is well aware that such revision is much needed, and trusts that his life and health may be spared to the completion of his ultimate design.

A word in reference to the doses of the concentrated remedies. Complaints have reached the author that of some of the preparations the doses indicated were too large, as of the Gelsemin, for instance. In the course of the work the writer frequently referred to the fact, that the doses named were such as he employed in the locality where he then resided, and that while the *properties* of the remedies would remain the same under all circumstances, the judgment of the practitioner must decide the propriety of *quantity, repetition,* and *continuance.* Since the present work was written, the author has had several months' experience in practice in the South, and has practically tested the fact, that in warm climates the doses of sedatives, narcotics, relaxants, &c., require to be diminished from 25 to 50 per cent., while stimulants and tonics require a proportionate increase. For instance, in the locality where the present lines are written, one-fourth of a grain of Gelsemin is equivalent in effect to one-half grain in the latitude of New York. On the contrary, three grains of Podophyllin are required here where two would answer the purpose at the North. The properties and employment of a remedy being given, it remains for the practitioner to graduate the dose. This can never be stated with such precision as to meet the necessities of every case, but only approximatively, time, circumstance, and idiosyncrasy forming the standard by which to judge.

Again this little volume is committed to the profession, with the assurance that the author holds himself strictly responsible for whatever of error as well as of truth may be incorporated in its pages, frankly inviting clinical criticism of all he may have said in relation to the properties and employment of the remedies considered. None are claimed to be *specifics* in the cure of disease, but all are claimed to possess *specific properties*, manifested, however, not uniformly, but in *specific* conditions. The condition ascertained, and adaptation of a suitable remedy being had, a manifestation of its specific powers may reasonably be expected.

That those of the profession into whose hands this volume may fall, will receive and test the opinions and statements put forth by the author, and render their verdict in the same spirit in which it was indicted, is the wish of

Their obedient servant,

GROVER COE.

Wilmington, N. C., June, 1860.

# CHAPTER I.

### CRUDE ORGANIC REMEDIES.

Objections to their Use—Of Uncertain Value—Frequently Inert—
Facts of Analysis, etc.

THE essential pre-requisites to the successful employment of Organic Remedies, are the possession of *specific therapeutic powers, uniformity of strength, non-liability to deteriorate by age, and convenience of administration.* Such remedies, either simple or compound, may be appropriately termed *positive medical agents. Positive,* not because they will *infallibly cure disease,* but because their sensible properties are *definite, uniform,* and *certain.* Such are the remedies of which it is proposed to treat in the present volume. In order to demonstrate the correctness of this appellation, it will be necessary to point out the deficiencies of crude medicines, and, by contrast, make apparent the superior claims of concentrated remedies to our confidence. To this end we shall endeavor to adduce a few facts in support of the exceptions we have taken to the use of crude remedies at the head of this chapter.

We have charged that they are of uncertain remedial value. By this we mean to be understood, that plants of the same species vary infinitely as to the amount of proximate principles inherent in them. The *fact* has been amply demonstrated by analysis. The *causes* we will endeavor to explain. Vegetable organisms may be said to be possessed of two constitutions, *physical* and *therapeutical*, blended into one *system*. By the term physical we would designate the structural apparatus of the plant; and the therapeutic to consist of the various secretions of this apparatus. By drawing a nicer line of distinction, we may divide the products of this apparatus into nutritive and medicinal. The constituents of this apparatus we term lignin, liber, &c.; of the nutritive products, amylum, gluten, sugar, mucilage, and albumen constitute the principal; while therapeutic constituents are variously denominated resins, resinoids, gum-resins, balsams, oils, alkaloids, neutrals, camphors, &c. In order to ensure the perfect development of the plant, it is evident that certain conditions of soil, climate, season, &c., must be present. The soil must be supplied with the various inorganic elements of the plant, and afford a sufficiency of water, in order that a proper degree of diluency of the various juices may be maintained. The climate must be such as will afford the requisite temperature, while the season must be of sufficient length to enable the plant to complete its numerous processes, and perfect its various parts. Any departure from these conditions will be followed by a corresponding deviation in the constituents of the plant. Poverty of the soil will starve the plant of its necessary food. Too high a temperature will urge on the various functions of the plant to complete its labors prematurely. Too low a temperature will retard the organic energies of the plant, and prolong its labors into the frosts of winter, which there will shut out all further chances of maturity. Excess of moisture, accompanied, as it must necessarily be, with a corresponding deficiency of sunlight and warmth, will exercise a strong influence over the future history of the plant. The burning sun of the summer drought, will, with insatiable thirst, drink dry every pore of the yielding soil, and the thirsting plants will droop and wither on the parched bosom of the parent earth. Thus do we behold the inevitable results which attend the working of nature's laws. Adaptation is the law of the universe, and in no light is it more vividly portrayed, than in its relation to the growth and development of the vegetable world. There are sermons in stones, and books in running brooks, saith the proverb. The vegetable kingdom may be called the very *printing press* of nature, each verdant leaf a type that prints a thousand varied impressions upon the quickened tablets of the reverent mind. The "still small voice" of creative wisdom is audible in all of nature's works, but the voiceless language of plants speaks most unassumingly in praise of "Him who doeth all things well." No study is more instructive, and at the same time interesting, than that of the laws which govern organic growth; and none more conducive to our best interest in this life. As the creatures of those laws, we must, of necessity, understand them, that we may be enabled to yield the allegiance implied in their establishment. By studying the physiology of plants, then, we may derive much instruction for the proper government of our own bodies. The facts set forth above in relation to the causes which influence the growth and development of plants, may teach us a useful lesson in regard to the conditions necessary to preserve the integrity of our own systems.

Dependent upon the causes above enumerated, plants are oftentimes entirely inert, so far as regards the possession of any therapeutic power. The vicissitudes of the climate and season may have so interrupted or suspended the secretive functions of the plant, that not a single proximate principle has been perfected. On the other hand, the absence of proper elements in the soil may have been the sole cause of the defect. Other causes might be enumerated, chief amongst which is, the gathering of a plant at an improper season. By so doing, the development of the proximate principles is arrested while they are yet, so to speak, in a transition state. The elaborating processes of the plant are arrested, perhaps, at the very moment when the various medicinal constituents are approximating the perfected principle. In such an event it is most certain that nothing of therapeutic value can attach to the plant.

In order that the reader may more fully understand our

meaning, we will endeavor to be more explicit. For this purpose, we will enter briefly into the physiological history of plants. To illustrate the subject, we will choose a perennial, deciduous plant of the temperate zone. The life of such a plant may be said to consist of an indefinite number of completed cycles periodically conjoined. These cycles are marked by four eras, spring, summer, autumn, and winter. During the winter months the organic energies of the plant lie dormant, nor are they awakened from their hibernal slumber except by the dawning of the succeeding era. This period illustrates the *static* condition of organic activity most forcibly. It is emphatically the season of rest, and may be appropriately termed the sleep of plants. Presuming that the labors of the previous seasons have completed the object of their mission, it is philosophical to suppose the plant to be complete in all its parts. Gathered at this season, and subjected to the searching powers of analysis, the manipulative skill of the chemist will penetrate each well-stored cell, and bring from their secret hiding places the various constituents of the organic body. Isolated, they stand forth as fractional representatives of the different constitutions of the plant—elementary parts of a compound system. This is the proper season to select such a plant, in order to determine its chemical constituency. This is the proper season to gather it for medicinal use; and this the season to collect it as timber for the purposes of the builder. Let us note the changes which follow an awakening of its latent forces. Now it may be compared to a well-stocked storehouse, wherein all the rich harvestings of the previous season are carefully laid by for future use. Let us watch how the present store may add in turn to the capital stock. The snows have melted under the thermal breath of returning spring, and gone to swell the volume of the turbid streams. The rigid, frozen earth has thrown off the icy chains that bound it in the embrace of winter, and its bosom swells with grateful pride as it drinks in the rich inspiring draughts of warm sunlight. The gentle showers descend, and the quickened soil presents, in each liberated pore, a willing reservoir. The time has now arrived for the resumption of organic activity on the part of the plant. The conditions necessary to this manifestation are, the presence of certain external stimuli. These consist of certain nutritious matters contained in the soil, water, atmospheric gases, electricity, and its allotropic conditions, light and heat. The stimuli of the soil are first available, being rendered so by the presence of water, and impelled by the electrical forces. The nutritive elements of the soil consist of carbon, silex, magnesia, lime, soda, potass, sulphur, the oxides of iron, alumina, etc. Water is the necessary vehicle of the nutritive elements of plants; but it is also decomposed, and its components, hydrogen and oxygen, enter into the structure of the plant. Carbon is also derived from the atmosphere, in the form of carbonic acid gas. The other elements afforded by the atmosphere, are, oxygen, both in its combined and simple form, and nitrogen. Before the nutritive matters of the soil can be appropriated by the plant, it is necessary that they should be in a state of solution. This is mainly effected by water. The roots, by means of minute vessels attached to their extremities, termed spongioles, now absorb the juices from the moist soil, and these, ascending, mingle with those already in the stem. These juices constitute the sap, so-called, of the plant. It holds, in solution, the proper nutritious substances which go to add to the volume of the plant, and also affords the necessary material for the reparation of its expended fluids. That it deposits some of its nutritive materials in its ascent, is undoubtedly true; but, of necessity, a certain portion must be conveyed to the extremities of its branches, in order that the gemmules, or buds, may receive the food necessary to their development, and the formation of leaves. The sap, in its ascent, has performed certain changes in the constituent principles of the plant, which, in the elucidation of our subject, it is important for us to notice. It has dissolved out a greater portion of the contents of the living cell, reduced them to a condition of solubility, and commingled them in one heterogenous mass. This we hold to be the established advent of the first era in the annual history of the plant.

Out of the ascending sap each part of the plant absorbs the material requisite for its nourishment. Following it in its ascent, we shall find that, as it successively reaches the buds, they swell, expand, and develope into leaves and flowers. The development of the leaves gives rise to the establishment of new functions on the part of the plant, which now will play a conspicuous part in its future history. Taking from the sap such materials as are necessary to the completion of this structure, they combine them with others drawn from the atmosphere, and appropriate the perfected constituents to the completion of their own apparatus. They are now in a condition to perform their share of the labor imposed by the establishment of organic activity, and to assist in the consummation of its object. Leaves have been denominated the lungs of plants. The similitude is correct, so far as regards the object, which is mutual, but will not apply to their functions. While the office of the lung is to absorb oxygen and give off carbonic acid, that of the leaf is to absorb carbonic acid and give off oxygen. In both instances the object is the preparation of nutritive materials for the purpose of organic growth and reparation.

We now have a period of organic activity which, at its culmination, will complete the first era. This is the final elaboration in the leaf of the various nutritive elements drawn from the soil and atmosphere, and their descent into the permanent structure of the plant. The fluid which, in its ascent, was called sap, has now, by its elaboration in the leaf, been converted into what is termed the proper juice of the plant. It is a highly elaborated, viscid fluid, composed of various rudimentary compounds, which, when reduced to perfected principles, will be recognised as starch, gluten, sugar, resins, gum, oils, alkaloids, resinoids, etc. The first era closes with what might properly be termed the completion of the digestive processes of the plant. The second era will comprise the period during which the nutritive apparatus of the plant makes appropriation of the duly elaborated materials. During this period the plant more sensibly increases in volume, new repositories are formed, and new stores laid in for a future season.

We would not be understood to imply that this is exclusively the period for the manifestation of these changes. On the contrary, we distinctly state, that these various phenomena are being carried forward during the entire period of organic activity. But we wish simply to impress the fact, that this is essentially the period when the organic stimulus is in its greatest force. During this, the second era, is the proper season for gathering leaves for medicinal use. They are now charged with the proper juice of the plant in a highly elaborate form. Should we wait until after the descent of the proper juice into the stem of the plant, we shall find that nothing but an exhausted apparatus is left behind. True, the leaf may preserve all its outward semblances of vitality, yet shall we find on analysis, that the therapeutic constituents are mostly wanting. The cellular tissue will be found deserted of its nutritive and medicinal substances, and their presence partially replaced with air A tree cut down during the second era, will be found useless for all the purposes of timber. The vital forces being mainly distributed to the periphery, that is, to the leaves, together with a greater portion of the vital constituents of the plant, the stem will be found to be deprived of too great a proportion of the preservative principles to enable it to resist decay. The alburnum commences a rapid decomposition, giving rise to a generation of worms, which, in turn, eat into the duramen or heart, and thus complete the destruction of the stem. We are assured by a gentleman from North Carolina, that a stem cut from a pine tree in the month of May, and placed in contact with the trunk of a healthy growing pine, will destroy it in the course of the season. The worms generating in the severed stem will pass to the living tree, and rapidly compass its destruction. We have seen the monarch of our northern forests, the lordly oak, when felled in June, pass into a state of complete decay in a space of from four to eight weeks.

Botanists have remarked that a plant early stripped of its leaves will soon perish. The reason given for this result is, that the absorption by the roots is insufficient to supply all the materials for its nourishment. This we have reason to believe, however, is not the sole cause. A great proportion of the resident nutritive materials of the plant having been dissolved out of the stem by the ascending sap, and carried in a state of solution to the leaves, it follows that if they are stripped off at this period, the stem will be exhausted beyond all chances of recuperation. A major part of the vitality of the plant is now at its circumference, and the severance of the leaves at this juncture will result in the hopeless impoverishment of the stem.

The third era in the annual history of the plant, comprises the period during which the products of the labors of the previous era are stored away in the various repositories of the stem; the withdrawal of the organic forces from the periphery; the exhalation of superfluous moisture; the fall of the leaf, and the suspension of all organic activity, preparatory to the coming of the fourth era, winter. Now, for a season, is all manifestation of organic activity withheld, and thus we have the completion of the cycle.

Let us recapitulate briefly the different stages of organic growth. First we have the ascending sap dissolving out the nutritive deposits of the root and stem, and conveying them to assist in the development of leaves and flowers. Now it is evident that if the root, bark, or stem of the plant be gathered at this season for medicinal purposes, it must, of necessity, be deficient of the constituents of which we are in pursuit. Not only will they be deficient in amount, but defective in composition; for, in arder to be of assimilative utility, the various constituents must be reduced to their rudimentary forms. Researches upon this point have established this fact beyond a doubt. Analysis has determined that the entire secretions of the living cells of the plant undergo complete disintegration and re-assimilation. What wonder then, if the plant be collected at this season, that we find it nearly or quite inert. If, on the other hand, we gather the leaves at this period, we shall find that they are premature and worthless. Nor, if we wait until the advent of the second era, shall we find that either the bark, root, or stem is of full therapeutic value. True, the proper juice is now descending, a new layer of cambium is being deposited, and the various parts of both liber and stem are succulent with the returning fluid. But much labor remains to be done ere the various proximate principles shall have reached organic completeness. The descending fluid is a heterogeneous mass, holding in solution the variously constituted compounds which go to replenish the various repositories of nutritive and medicinal substances. The absorbent and assimilative powers of the plant are now directed to this mass, its constituents isolated, taken up and deposited in their appropriate receptacles. Although winter is the period when we should look for the highest degree of perfection in the medicinal principles of a living plant, such as we have described, yet we cannot say, with truthfulness, that the cessation of all outward manifestations of organic activity argues perfectability in the various constituents of the plant. The labor of assimilation is still going on within its silent organism. The precise moment when this assimilative action has reached its highest point of culmination is very difficult to determine, even in the living plant. How much more so, then, in the detached portions of the dead specimen. If, during the life of the plant, organic activity has done its complete work, then may we expect that the death of the plant will usher in a period during which material changes will be effected in its constituents, terminating only by their reduction to primary forms, or entrance into new combinations. The laws of chemical decomposition and recombination know no rest. Their action is as ceaseless as the footsteps of time. All created matters feel their mighty impress, and yield resistless to the eternal law of mutation.

The peculiar chemical action which goes on in the constituents of dried plants, is productive of directly opposite results. In the one case it tends to perfect, or we should say rather, to render available certain peculiar principles. We have an example in the concrete juice of the Fraxinus Ornus or manna ash—the manna of commerce. This substance increases in purgative qualities by age. Some reaction of its constituents upon each other undoubtedly produces this result. One of the principal constituents of manna is mucilage, known by its yielding mucic acid. It is not strictly a proximate principle, but contains bassorin, cerasin, &c. This substance acting upon the nitrogenous constituents of the manna, effects their decomposition, brings about new combinations, and thereby increases its purgative power.

The oak bark employed in tanning leather improves in value for a period of four or five years after it is stripped from the stem. So well established is this fact, that, where capital will permit, a stock is constantly kept from two to five years ahead. The reason of this we will now explain. Tannic acid cannot properly be considered a proximate principle of vegetable organisms. It never exists in the *living cells* of the plant, but is the legitimate product of a peculiar putrefactive decomposition which takes place in the *dead cells*. Proximate principles are those which undergo progressive formation in the living cells of the plant during the period of organic activity. But tannic acid is the result of a regressive chemical action within the dead cells. As it is found only in the dead cells of the living plant, it follows that the arresting of the life of the plant will, by destroying the vitality of the cells, favor the decomposition which results in the formation of this principle. We shall have occasion to revert to this subject in the next chapter. It is in this way that age augments the amount of tannic acid in the bark, and gives to it increased value.

The Rubia Tinctorium, a root much in use by dyers, improves in value for an equal number of years. It is never employed until it has attained the age of two years, dating from the period of its collection. Here, again, certain chemical decompositions take place in the interior structure of the plant, which give rise to new combinations, whereby the peculiar principle for which the plant is esteemed is largely increased in amount.

Apples, pears, peaches, oranges, and other fruits, undergo a series of ripening processes after they are detached from the plant that bore them. The peculiar action here involved, is the conversion of starch into sugar, and the development of the flavoring principle. Coffee so improves in flavor by age, that the most inferior kinds are said to rival the finest Mocha, after having been kept for a period of from ten to fourteen years. Tobacco is also subject to the same improvement. Instances might be multiplied, but we deem the above sufficient for illustration.

Thus we see, that even after the continuity of the different parts of the plant is broken up, the detached portions are silently, yet surely, undergoing important constitutional changes. In the cases above cited, this peculiar action tends towards desirable results. But we shall see that age is equally potent in the destruction of the perfected proximate principles of the dried plant. These changes, as we shall show, render it valueless. While the plant is endowed with organic life, it possesses the power of resisting the action of external disintegrating influences. But, when deprived of that life, it becomes a prey to those active disorganising agents, air and moisture. Indeed, within its own substance it conceals those restless agencies which are instrumental in effecting the dissolution of vegetable organisms.

External appearances, it will be shown, do not afford reliable indications of the therapeutic value of plants. Therefore, the presence or absence of proximate medicinal principles cannot be ascertained by visual scrutiny. Neither the giving off by the plant of its natural odor, nor the preservation of its peculiar color, can be relied upon as evidence of therapeutic worth. The flavoring and coloring matters, although of medicinal value, are distinct principles, and may exist independent of the more active medicinal constituents. Hence no degree of certainty can attach to outward signs. A quantitive analysis alone, by isolating its various constituents, can determine the fact of the presence or absence of the inherent proximate medicinal principles of any given plant. Though perfect when collected by the botanist, time may have effected the reduction and dissipation of its constituents, or rendered them into new combinations. In the one instance they are made valueless;

and in the other, their character is changed, and rendered uncertain. On the other hand, climatic, meteoric, and other influences, separately or combined, may have effectually prevented organic completeness, by arresting the growth of the plant ere maturity.

Winter, then, is apparently the season for collecting such a plant as we have described for medicinal purposes. We would naturally expect to find in such a plant, at this season, an entire completeness in its organism. The reader will perceive, from the facts above set forth, that the directions given by some botanists for collecting barks in the season when they will peel from the stem are erroneous.

We hope we have now made it apparent to all how liable plants are to suffer from the vicissitudes of soil, climate, season, &c.; and how liable they are to vary as to the amount of the various proximate principles attributed to them. Repeated analysis have demonstrated the fact, that specimens of the same plant grown in different localities will vary infinitely in the proportions of active principles yielded. The want of a knowledge of this fact has given rise to much contrariety of sentiment amongst practitioners in different sections of the country, in regard to the remedial value of various plants. The Scutellaria Lateriflora has been condemned by some practitioners as inert and worthless, while others set a high estimate upon its value as a nervine tonic. It remains for organic chemistry to reconcile this difference of opinion. Analysis of various samples of this plant grown in different sections of the eastern States, has proven it to be very deficient in the active principles attributed to it. The yield of various samples, amounting in the aggregate to over one thousand pounds, was not sufficient to pay the first cost of materials. On the other hand, samples of the same plant grown at the South and West have yielded a fair proportion of the proximate principles belonging to the plant. The Senecio Gracilis varies remarkably as to its yield of active principles. From the analysis of a great number of samples, at different times, it has been found that the yield from a given quantity will vary in the proportions of from one to four.

With the Helonias Dioica the same variableness has been found. In this plant the variations have been remarked to be from two to five. The plants, in every instance, bore upon their exterior an equally promising aspect. Analysis alone could detect and make apparent the deficiency. Here is a discrepancy which can be accounted for only upon the grounds we have above shadowed forth.

The Asclepias Tuberosa, growing in the comparatively barren and sandy soil of New Jersey, yields from one to two hundred per cent. more of Asclepin, than that grown in the rich alluvions of the West. Numerous other plants might be mentioned, the analyses of which have been attended with like results; but we deem these sufficient to illustrate the fact. From this it will be seen that uniformity of therapeutic power can never be looked for in crude remedies. Suppose we take, for instance, the usual formulas of the dispensatories for the preparation of infusions and decoctions. A given amount, by weight, of some root, bark, or herb, is directed to be added to a stipulated quantity of water. The dose is defined, and the necessary requisitions are considered complete. Now let us look a moment at the reliability of such a preparation. Bearing in mind the facts previously adduced, the reader will easily follow us to a common conclusion. Water being the menstruum, it follows that the active principles it is capable of holding in solution, can not be other than neutrals, alkaloids, acids, gums, mucilages, and coloring matters. Now what guarantee have we of the value of such infusion or decoction? We have seen that plants bearing on their exterior all the marks of genuineness, have, on analysis, been found nearly destitute of any medicinal principles whatever.

Admitting that the plant has been grown under the most favorable auspices, we yet shall see that the actual amount of active principles present will be indefinite. No two samples of the same plant yet analysed have given a uniform amount of proximate principles, no matter how favorable the conditions accompanying their growth. Considering the liability of plants, then, to be influenced in their development by the vicissitudes heretofore enumerated, it will readily be perceived how much more indefinite must be the remedial value of a plant, the circumstances of the growth of which we know nothing. We are informed by the dispensatory that the amount of a certain alkaloid (*Narcotin*,) afforded by even the same varities of opium, will vary from 1.30 to 9.36 per cent. This discrepancy amounts to over 700 per cent., and, with so potent a remedy, is a matter of great moment.* Even admitting a uniformity of constitution in the article employed, we yet shall see that but a short time is necessary to effect a complete decomposition of the therapeutic constituents. We are further told by the authority above quoted that certain decoctions and infusions, in warm weather, "speedily run into the putrefactive fermentation." The philosophy of this peculiar decomposition we shall explain in a future chapter. We refer to it at the present time only as an additional objection to the employment, or rather method of preparation, of crude medicines. The neutral principle of plants is that which is most liable to be decomposed by this peculiar chemical decomposition, and yet it is the principal constituent usually afforded in aqueous preparations. From either, or a combination of the causes we have enumerated, practitioners have, no doubt, been frequently disappointed in the anticipated remedial value of watery preparations. If we should ask, what reliance can be placed upon preparations so uncertain in therapeutic

---

* This is also true as regards the yield of *morphia*. Sometimes this alkaloid is almost entirely wanting. We are informed by an eminent physician of this city, that a friend of his lately returned from India, states that in a wet season, although the product of opium is increased, yet it is found to be almost entirely deficient of *morphia*. A dry, hot season seems to be most favorable to the production of this principle. Suit was brought in this city a few years since for the recovery of the value of 8000 pounds of opium, which had been purchased for the purpose of manufacturing morphine. On analysis the drug was found to be nearly destitute of this alkaloid. Hence, from a want of a knowledge of the true cause, a charge of fraud was preferred. The Cannabis Indica grown upon the elevated ridges of India is extremely different from that grown in the vallies. Locality, as well as other circumstances of growth, seems to wield a potent influence in the development of medicinal plants.

strength, it might be answered, that the physician will determine their utility by experimental administration. True, by such a course, their comparative value might be ascertained, but are not such experiments extremely hazardous, both to the interests of patient and physician? In urgent cases, *time* is of the greatest moment, and its lavish expenditure in instituting a series of clinical qualitive and quantitive analyses under such circumstances, in order to test the therapeutic value of any given remedy, could scarcely be looked upon in any other light than criminal. In its most favorable aspect, a degree of recklessness would attach to it which no conscientious physician would willingly countenance.

Nor if we employ the remedy in substance, shall we have arrived at any greater degree of exactitude. If we write a prescription for a pill of crude opium, how shall we, by the above showing, be enabled to tell anything of the proportions of the nineteen or twenty constituent principles attributed to it? As the amount of some of its most active constituents vary from 1.80 to 9.86 per cent. in a given quantity, it is apparent that great uncertainty must attend its exhibition. These facts admit of a wide range of application. Such of the medicinal plants as contain highly active constituents, for instance the Digitalis, and others of its class, can never be understandingly exhibited, either in infusion, tincture, or substance. The same may be said of all crude organic remedies, but more nearly concerns those possessed of peculiar potency. Morphine is a positive medical agent, being of definite, uniform and certain power. Not so with Opium. Here the therapeutic constituents are blended with, and diffused through, a mass of non-medicinal substances, the number and amount of which can only be determined by analysis. Here it is that the scrutinising powers of organic chemistry display their peculiar utility. Divesting the therapeutic constituents of all extraneous admixture, it hands them over to the physician, defined in amount, character, and sensible properties. This fits them with those characteristics which enable the practitioner to understandingly and successfully employ them.

Another division of the influences which have a bearing upon the history of organic remedies, now claims our consideration; and that is, the artificial cultivation of medicinal plants. In the transference of plants from their native localities to soils prepared by the hand of man, many and important changes are effected in their individual constitutions. This is an established fact in regard to vegetables used as food, which has long been recognised by botanists; but we are not aware that the subject, as it relates to the changes effected in medicinal plants, has been so fully elucidated. The natural order of Crucifera, tribe Brassica, furnishes many examples of plants reclaimed from their wild habitudes, and rendered subservient to the purposes of food. True, all esculents must have been domesticated by the genius of man at some period of the world's history, but the greater number of them date the advent of their initial culture so remotely, that we have little information respecting their primeval habits or characters. Of those above referred to, botanists have been enabled to note the changes effected by cultivation. Many plants now cultivated for the table, were formerly esteemed exclusively as medicines. Cultivation has converted the small acrid root of the *brassica rapa*, or turnip, into a large and nutritious article of diet. Numerous similar illustrations might be adduced, but we presume our readers are already familiar with the facts. Now if plants can be so essentially changed in their characteristics that, from being bitter, acrid, and worthless as food, they become nutritious, palatable and wholesome, we have but to transfer the application of the principle to medicinal plants reared in the garden of the botanist, to see that our exceptions to the artificial cultivation of medicinal plants are well taken. By such a procedure they are much deteriorated in medicinal value, and often rendered entirely worthless. Take, for example, the *Leontodon Taraxicum*, or dandelion. That which is grown in natural localities possesses well defined and efficient remedial powers. True, much controversy has been had in relation to its therapeutic worth, and much been said, both *pro* and *con*. Much has been said and written to prove its inutility, and with many practitioners it has fallen into disrepute. But the reason for this, as we shall show in a future chapter, when treating of extracts, has not always been the natural defects of the plant, but of the method of its preparation.

By instituting a comparison between the dandelion of the shops—we mean such as has been artificially cultivated—and that collected from its native haunts, many important differences will be found, not only in its external aspect, but also in its analytical and therapeutic peculiarities. In the cultivated plant the proportions of starch, grape-sugar, and other non-medicinal constituents are largely increased; while the amount of proximate principles is proportionably diminished. Medicinally, the native plant is of well established utility in the treatment of a variety of diseases, particularly affections of the liver, kidneys, and respiratory system. Let any practitioner skeptical of its remedial value, gather the plant in the month of August, express the juice, and administer it in table-spoonful doses to such as are laboring under hepatic derangement, and he will fully realise the fact of its power to produce decided and sanative physiological results. That this is true of the recent plant, admits of no doubt; but the great difficulty consists in so curing or pharmaceutically preparing the plant as to preserve its peculiar virtues. The process of kiln-drying medicinal plants is another most objectionable feature in the history of such as are artificially reared. By this process the volatile principles are dissipated, and certain chemical changes effected in other of the constituents. We need not multiply instances to make the fact, that material changes are effected in the constituents of medicinal plants, by artificial culture, patent to the mind of the reader. That even the structural aspect of plants may be altered by cultivation, is illustrated in the case of the Rose, in which, by culture, the stamens have been converted into petals.

Plants also adapt their habits to the circumstances under which they are placed. The evergreens of the south become deciduous when transplanted to a northern clime. For example, the Magnolia Grandiflora, and others. The Castor Oil plant, which in Africa forms woody trees, becomes an annual in our gardens. The Mignonette, which, in Europe, is an annual plant, becomes perennial in the sandy deserts of Egypt. Thus, on either hand, do plants conform their habitudes to the circumstances of their exposure. If, then, as we have seen, plants can so essentially change in their habits and external forms, is it not reasonable to suppose that they are capable of being materially altered in the chemistry of their organism. But we do not have to depend upon supposition in the latter instance more than in the former. We have the corroborative tests of analysis to sustain our inferences of the fact.

While we wish to adhere to our advocacy of the fact, that cultivation materially affects the therapeutic constituency of plants, we do not wish to be understood to imply that said fact invariably militates against their comparative value. On the contrary, we are aware that cultivation has had much to do in developing and augmenting the medicinal as well as the nutritive value of certain plants. Their number, however, is comparatively few. We might mention the Poppy, Hops, and various species of Labiatæ which yield the aromatic oils of commerce. Success in these instances, however, depends upon accident of adaptation. Soil, climate, season, exposure, all unite in conducing to this end, or conspire in militating against the perfect development of the plant. We are of opinion that very little attention has been given to the question of adaptation in all its essential requisites, and that chance alone has favored the experiments. In this opinion we are confirmed by the perusal of all the treatises upon the artificial rearing of medicinal plants to which we have had access. Not only is no mention made of the chemical qualities of the soil, exposure, length of season required for development, etc.; but seldom are the chemical constituents of the plant defined with anything like precision. These omissions seem peculiarly pertinent to the question of the successful cultivation of medicinal plants. Attempts have been made, in England, to cultivate the Rhubarb for medical purposes, but popular predilection so much favored the imported root, that the project has been nearly or quite abandoned. The preference, in this instance, was based upon the accredited superiority of the foreign article, while a consideration of the essential causes of the difference have no share in the formation of the opinion. Clinical experiment demonstrated the relative value of the two, and here the question rested. The fact seems not to have incited a very rigid inquiry into the philosophy.

We hope that we have now established the various points of our argument. Inasmuch as we have demonstrated the fact, that plants vary infinitely in regard to the amount of active principles yielded by different samples of the same species; that the vicissitudes of soil, climate, season, exposure, &c., all conspire in influencing the growth and development of the plant—that the period of collecting, and method of curing exercise great control over the constituency and preservation of its active principles—the external appearances are no indication of reliability—that cultivation changes, and renders uncertain its essential therapeutic properties—and that by age the medicinal constituents of the dried plant are decomposed and dissipated, we hold that the exceptions at the head of this chapter were well taken. We have shown that crude organic remedies can never be of definite, much less of uniform therapeutic power. These points, setting aside all consideration of the causes, have been amply demonstrated by analysis. That they are frequently inert, has been substantiated by the same authority. These facts alone are sufficient to prove them non-reliable, and, at best, of uncertain value. It follows then, that no matter what form we may exhibit them in, we will not arrive at any degree of definiteness in regard to their remedial value. Be it in substance, tincture, infusion, decoction, syrup or extract, the same uncertainty will ever be attendant. Experiment alone can determine the relative value of each preparation; but to such a proceeding, in the present state of pharmaceutical science, attaches a high degree of culpability. A knowledge of the facts set forth in this chapter being accessible to all who desire to learn, no excuse can be accepted from any one for not availing himself of the superior advantages offered by concentrated medicines.

We are far from advising that the ordinary methods for the preparation of the simpler plants should be abandoned. On the contrary, we are a strong advocate for the employment of the simpler vegetable agents as auxiliaries in the treatment of disease. In our own practice we make frequent use of such agencies, in infusion, decoction, etc. But we confine ourselves to such incidental plants as may not yet have been prepared in a concentrated form and whose properties are such as not to render their indefinite administration hazardous. But with all the more potent agencies, and where efficiency and promptitude of action is demanded, we have long ago dispensed with the employment of other than concentrated agents.

We now come to a consideration of the chemical properties of vegetable constituents, and the *rationale* of the reactions whereby the proximate principles are decomposed. To this subject we shall devote another chapter.

# CHAPTER I.

## CONSTITUENTS OF PLANTS.

Acids—Alkaloids—Indifferent or Neutral Substances, etc.

No branch of human knowledge is so much indebted to the researches and developments of the chemist, as that of the science of medicine. He it is who prepares and provides the physician with means wherewith to do battle against the many "ills to which flesh is heir." He defines the laws which govern the form, properties, and affinities of matter, thus furnishing the physician with a chart to guide him safely o'er the troubled sea of medical practice. Even the physician himself must become a chemist—a chemist of the higher order of organic chemistry. His duty it is to control the chemical processes of life; to harmonise irregularities and correct morbid conditions by means of reagents. It devolves upon him to superintend the formation, secretion and excretion of chemical combinations. It is necessary therefore, that he should be acquainted with the laws which govern chemical action, and with the properties of the reagents he employs. He must understand what particular circumstances and external influences will diminish, or completely suppress the efficacy of his reagents. He must know whether his reagents will radically cure disease, or whether they will simply afford temporary relief, entailing still greater complications by their reaction. He must know whether they will relieve a lesser evil by the substitution of a greater;

whether the substances conveyed into the system are capable of healthful assimilation; or whether they will form combinations destructive to the integrity of animal organisms. Provided with this knowledge, he will be enabled to practice his profession understandingly and successfully.

In order to a better understanding of the remedies treated of in this volume, we now propose to consider, in detail, the various proximate principles of which they are composed. To do this more comprehensively, we will first consider each of the principles separately, defining their sensible properties in the isolated form, and finally treat of them in a state of combination.

The number of single substances produced by vegetable activity is very great. Many of these substances are very little understood, if, indeed, they are known at all. Certain substances are common to all plants, and constitute the materials of vegetable formations. These, by way of distinction, we term nutritive. Again, there are substances which are found only in a certain class of plants; while others are peculiar to a single plant. Upon the peculiar properties of these substances is based their employment in medicine. Such are designated therapeutic principles.

In considering the chemical properties of vegetable substances, we will divide them into the three following classes:

CLASS I.—VEGETABLE ACIDS.

In the strong affinity displayed by these substances for bases, they much resemble the inorganic acids. With few exceptions, they are crystallizable and soluble in water. The greater number of them yield crystalline salts with bases. Nearly all vegetable acids change the color of blue litmus paper to red. Some of these acids are common to a large number of plants; others are found only in a certain genera; while some are confined to a single species. A part of the acids common to plants are the products of organic growth, while others are formed only after the vital activity of the plant has ceased. These latter are formed by the decomposition of the constituents of the plant, under the agency of external influences. The vegetable acids are mostly found in the nutritive constituents of plants, and but few of them possess any peculiar medicinal value. They exist partly free, partly united with bases, and partly in combination with neutral substances.

Vegetable acids are formed by the conversion of amylum and oil into cell-substance. By the operation of the same vital power, acids are again reduced and reassimilated to the primary form. Therefore if we find in plants a peculiar oil, common only to a certain class, or an individual species, we may be sure to find, also, a peculiar acid; and if the oil possess therapeutic value, the acid will possess it likewise, although in a modified form.

A large number of acid principles are employed in medicine which depend for therapeutic value upon their astringent properties. These form a class to which we shall give the name of *tanneous* acids. These acids are not, as we have stated in the preceding chapter, strictly speaking, proximate principles; that is, they are not formed in the living cells of the plant during the season of organic activity, but are the product of a peculiar putrefactive decomposition which takes place in the dead cells, whereby the cellulose is converted into tannin. As tannin is found only in the dead cells, and as cellulose is converted by the vital processes of organic activity into wood and cork substances only, it follows, therefore, that tannin is a product, not of organic formative power, but of regressive chemical action.

These tannin substances are not distinct principles, but are composed of a number of different principles combined together. A part of these substances, only, give acid reactions; that is change blue litmus paper to red. They are known by their astringent taste—by giving with the salts of iron, blue, black, and dirty green precipitates—and by their power of combining with animal skin. With protein substances they form insoluble compounds. This we hold to be a strong reason why tannin cannot exist in the living cell. It would combine with and coagulate the contents of the cell, and render the albuminous matters, those great reagents of vegetable activity, insolu ble. Thus would the nutritive constituents be rendered unavailable, the secretions checked, progressive formation arrested, and the functions of organic life suspended.

By the action of water and oxygen, tanneous acids are converted into a brown colored substance, but slightly soluble in water, termed *humus*. In this respect they resemble the extractive or neutral substances.

Tanneous acids exist in great variety, and of very different properties. Those derived from gallnuts, oak bark, &c., are distinguished by the name of *gallo-tannic, querci-tannic*, etc. Another class is derived from Peruvian bark, catechu, &c. And still another class belongs to the indifferent or neutral vegetable compounds, and gives to many astringent plants their principal medicinal value.

CLASS II.—ALKALOIDS, or *Vegetable bases.*

These are certain organic compounds, which, on account of their possessing properties analagous to inorganic bases, particularly alkalies, have received the above appellation. The greater number of these substances change red litmus paper blue. They all combine with acids, and form crystallizable salts, out of which they may be again separated by the action of a stronger base. It is from this similarity in their chemical reactions to the mineral alkalies, that they have received the name of alkaloids. The vegetable alkaloids exist both in a solid and liquid form. The former are mostly crystalizable, with few exceptions are colorless, non-volatile, and have but a faint odor. The latter are volatile, and have a stronger odor. By far the greater number of vegetable alkaloids have a bitter taste, and are more soluble in alcohol than in water; while a few, like the *Hyosciamine*, are more soluble in water than in alcohol.

Most vegetable alkaloids are composed of oxygen, hydrogen, carbon, and nitrogen. A few, the liquid alkaloids particularly, are composed of hydrogen, carbon, and nitrogen only; while we occasionally find an inorganic element, as in the *Thiosinnamine*, which consists of hydrogen, carbon, nitrogen and sulphur.

Alkaloids do not exist free in plants, but are generally combined with acids, forming acidulous, and but slightly soluble salts. They are formed by the reaction of the bast-cells upon their abuminous contents, or the so-called milk sap of the plant, and are produced only in living plants. They are solely the products of organic activity, and their quantity is never increased in the dead or dried plant. Plants are generally richest in alkaloids during the winter months; that is, after the cessation of the vegetating process, and while they are enjoying their hibernal sleep.

When plants are undergoing decomposition, alkaloids are the last of the medicinal principles to be attacked ; but they are liable to be greatly modified or completely subverted in therapeutic value by the products which arise from the decomposition of the constituents of the plant. For instance, if tannic acid be formed in considerable quantities, it will combine with and completely suppress their activity as remedial agents, the *bi-tannate* of every vegetable alkaloid being entirely insoluble in all menstrua except stronger acids. How important this knowledge is to the physician, that he may avoid combining together incompatable principles, and thereby render nugatory their medicinal power.

Some of the alkaloid principles of plants form their most active and valuable medicinal constituents, while others are possessed of but feeble properties.

CLASS III.—Indifferent or Neutral Principles.

This class embraces all the remaining substances of vegetable activity, and which are of very diverse chemical characteristics. A part of these are formed during the period of functional activity, and part are the result of subsequent decomposition. They are called neutral because they have not the power of neutralizing acids or bases, although they often combine with both. A larger number of these substances are more nearly allied to the class of acids, and are evidently of an electro-negative character. Amongst these may be enumerated a large number of resins, extractive or neutral principles, coloring matters, and products of decomposition.

Others partake more of the basic character, of which we have examples in the *ether* resulting from the decomposition of alcohol, and the *methyl-oxyd* obtained from wood spirit. Many of these substances, as the greater portion of the fats, fixed oils, and ethereal compounds, partake of the character of salts. Such of the volatile oils as contain no oxygen, may be considered as simply hydrogenous compounds of hydro-carbon radicals. The greater number of the principal substances of plants are oxydes, either with or without nitrogen, but without any distinct chemical character, being, as before stated, neither basic nor acid, yet possessing the power, under some circumstances, of combining with both. When submitted to chemical processes, they yield for the most part acid and basic substances.

There are many substances pertaining to plants, some of which are common to all plants, others are distributed through different genera, which possess no particular medicinal value, and which it might be thought unnecessary to notice in this connection. But their importance, in view of the results produced by their reaction upon the therapeutic constituents of the plant, demand that we should examine their history and influence more closely, before we enter upon a description of the properties and employment of the remedial agents under consideration. We will consider first—

CELLULOSE.—This substance is also known by the name of *cell-substance*, or *cell-membrane*. Its formula is $C_{12} H_{20} O_{10}$. It is a white flexible mass, without smell or taste, insoluble in water, alcohol, ether, alkalies and concentrated hydrochloric and nitric acids. Exposed to destructive dry distillation, it does not soften, nor melt, but is converted into a beautiful colorless charcoal, which retains the cell form. Under the microscope it has the transparency and appearance of diamond. Under the action of concentrated sulphuric acid it is dissolved and converted into *dextrine*, and by long continued boiling in water acidulated with sulphuric acid, it yields, like dextrine, crystalizable *grape-sugar*. A solution of caustic potassa simply causes it to swell, without producing any further visible change. A solution of iodine colors cellulose of a pale yellow; chloriodide of zinc gives, in some cases, a blue color, as also will iodine and sulphuric acid, but in other cases gives only a pale yellow, It seems evident, therefore, from this, that cellulose exists under different modifications. By the progressive organic activity of the plant, it is converted into *wood* and *cork substance;* while, by regressive chemical action, it is converted into *amylum* or *starch.*

XYLOGEN.— *Wood substance.*—This principle is but slightly soluble in concentrated sulphuric acid, but is easily and completely dissolved by caustic potassa; also, when boiled with chlorate of potassa and nitric acid. When boiled in dilute sulphuric acid for a long time, it yields *grape-sugar*, in which respect it is similar to *cellulose*. Chloriodide of zinc, and iodine and sulphuric acid do not color it blue. Xylogen is found in the primary cell-wall, and in the thickening layers of all woody cells. It is formed by the progressive conversion of cellulose, and gives stiffness to the cell-wall. Its quantitive elementary analysis has not yet been made.

*Cuticular* or *Cork-substance* is somewhat similar in its character to the preceding. It is not soluble in sulphuric acid, neither is it invariably completely soluble in a solution of caustic potassa. It differs from xylogen in its behavior towards oxydising agents. When boiled with chlorate of potassa and nitric acid, it is converted into a resinous substance, which is soluble in alcohol and ether, and burns with a strong, shining, but smoky flame, giving off a feeble aromatic odor, and leaving a porous coal.

Cuticular substance is found in the walls of the older cork-cells, which are frequently formed entirely of this material, and in the so-called cuticular layers of the cells of the epidermis. Cuticular substance is also the product of a progressive conversion of cellulose, which diminishes in quantity in proportion to the increase of the former, until, in the perfected cork formation, it has entirely disappeared. The chemical differences between cuticular substance and xylogen are produced by the direct reaction of air and light upon the former, to which it is more exposed. Cuticular substance has the power of preventing the diffusion of fluids. It prevents exhalation from the surface, and the commingling of the sap of neighboring cells. It is therefore a substance of great importance in the vegetable economy.

PROTEIN.—Protein substances are insoluble in concentrated sulphuric acid, but are dissolved by caustic potassa. Iodine and sulphuric acid color them of a light golden yellow. By the action of chloriodide of zinc they are coagulated. Hydrochloric acid gives, after twenty-four hours, a violet color. Sugar and sulphuric acid the same. Nitrate of mercury produces, after from five to ten minutes, a beautiful rose-colored tint.

Protein substances eagerly absorb soluble coloring matters. By taking advantage of this property, flowers may be artificially colored by mixing soluble coloring matters with the soil in which they grow. In this way, flowers which are naturally white, may be rendered blue, red, and even black. Flowers naturally colored may be made to partake of all intermediate hues.

A part of these protein substances, mixed with other matters, enter into the structure of the cell-walls, and the rest form a portion of the contents of the cell. This latter portion we term *protoplasma*. It is a granular, slimy, nitrogenous liquid, and is found collected upon the inner surface of the cell-wall, and surrounding the nucleus of the cell. It does not mix with the cell contents, but in young cells is frequently found suspended in the cell-sap, and travelling in currents. Protoplasma has been long know under various names, such as *vegetable mucous*, *vegetable glue*, etc.; but it is only by later and more strict investigation of its character and properties, that its importance in the vegetable economy has been fully established.

It constitutes the different degrees of development, degrees of oxydation we may say, of protein, the radical of albuminous substances, either combined with sulphur alone, or with sulphur and phosphorus. By their chemical properties, three groups of protein substances may be distinguished, viz; *albumen, legumin or casein*, and the so-called *fibrin*. This term, however, is a misnomer, as it bears but a slight similarity to animal fibrin, and is never identical with it. Fibrin is the product of a higher organic activity, originated solely in animal cells, and formed by the action of the life forces upon the constituents of animal fluidity.

Protein substances are found only in living cells, and as soon as the cell-sap is consumed, and the cell dies, these substances disappear, and the cell becomes filled with air. The current motion which we spoke of as being apparent in the protoplasma of the cells, is produced by a reciprocal chemical reaction which takes place between the protein substances and the rest of the cell contents. Protoplasma is of the greatest importance to the cells, constituting, as it were, their life agency. It assimilates the various substances brought into the cells, conduces to the formation of cell-substance, and its separation in the form of a membrane. The proper cell-sap is passive, while the protoplasma is circulating through it in never ceasing currents; and while the protoplasma is of itself undergoing continuous material changes, it effects a metamophosis in all the constituents of plants. Alkaloids, and all other nitrogenous medicinal principles of plants, are formed by the direct decomposition of these protein substances. They also afford the materials and stimulus for an increased production of their primary substances. These protein substances are readily decomposed by the action of heat, air, and water. This power of spontaneous decomposition is transferred to substances of a more constant and enduring composition, when brought in contact with them, which property renders them great promoters of fermentation and putrefaction. Protein substances exist in two conditions or degrees of modification, one class being soluble the other insoluble in water. In chemical properties they are very similar.

AMYLUM.—($C_{12}$ $H_{22}$ $O_{10}$)—Starch is the most common of all the substances concerned in vegetable activity, being found in all plants, although it is not everywhere present in the same plant at all seasons. It is principally the product of the parenchymous cells, but is likewise found in the cells of the medullary rays, and also in some of the bast-cells, of which we have examples in the various species of Euphorbiæ. Amylum is produced in the plant at certain periods of its growth, and at other periods is again consumed, entering into new combinations, and forming new products. It is seldom to be met with in an amorphous condition, but usually occurs in grains of different form and size, the smallest of which measure $\frac{2}{1000}$ of a millimetre, and the largest $\frac{185}{1000}$ of a millimetre in diameter. These grains are each one composed of numerous layers. They are colorless, transparent, and insoluble in cold water, alcohol and ether. When boiled with water, or treated with acids, they are simply swelled up and suspended, but are not really dissolved. Paste consists of these starch grains swelled up in water. When this paste is further diluted with water and filtered, the starch remains upon the filter, as is proven by the filtrate not being colored by a solution of iodine. Starch grains, both solid and swelled in water, are colored by iodine. The color varies from a light wine-red, to a deep indigo-blue, according to the amount of iodine present. In the production of this color the iodine has combined chemically with the starch, forming an iodide. Iodide of starch does not swell up either in sulphuric or hydrochloric acids, nor in boiling water, so long as free iodine is present, but remains unaltered; but as soon as the iodine is volatilised it is decomposed.

Dry starch is colored brown by iodine, but becomes instantly blue on the addition of water. Chloriodide of zinc colors starch of various shades, from violet to blue. When starch is swelled up in water and allowed to stand for a length of time, it undergoes successive chemical changes or decompositions. It is first converted into *dextrine*, then into *grape-sugar*, and finally into *acetic* and *lactic acids*. The decompositions are brought about by the reaction of the albuminous substances with which starch is always admixed. When boiled in diluted sulphuric acid, starch is gradually converted into *dextrine* and when the process is further continued, into crystallisable *grape-sugar*. Boiled in diluted nitric acid, *oxalic* and *mucic acids* are produced.

Starch is generally converted by the processes of vegetable activity into *dextrine* and *sugar*. Out of these products cellulose is formed by the reaction of the protoplasma and cell-nucleus; and by the further action of the protoplasma is separated in the form of membrane. On the other hand, cellulose is again reduced by the organic processes and re-converted into starch. It follows, as a matter of course, that these progressive and regressive metamorphoses must give rise to a number of substances, degrees of formation they might be termed, which, though isomeric in their composition, differ essentially in their physical and chemical properties. *Inulin* seems to be substituted in some plants for amylum. Its properties and manner of origin are similar, and in reality it is nothing else but amylum in a certain stage of transformation.

DEXTRINE.—($C_{12}$ $H_{20}$ $O_{10}$.)—This is a yellowish, or dark brown, lustrous, brittle substance, insoluble in alcohol and ether. Water dissolves it to a tasteless, frothy, viscous liquid. When boiled with dilute sulphuric acid, it yields *grape-sugar*. Iodine does not color it blue. When diastase is mixed with a solution of dextrine and allowed to stand, grape-sugar is formed. Dextrine occurs in plants dissolved in the cell-sap. It is produced by the transformation of amylum, and constitutes the intermediate stage between starch and sugar. Dextrine is active in the formation of new cells.

SUGARS.—Six varieties of sugar are known in chemistry, four of which are crystallisable, and two uncrystallisable. Of the crystallisable, we have *cane-sugar*, *mushroom-sugar*, *grape sugar* and *milk-sugar*. The uncrystallisable are the *fruit sugars*, and *treacle*. Milk-sugar occurs only in the milk of animals.

Cane-sugar has the formula, $C_{12}$ $H_{20}$ $O_{10}$. Grape-sugar, $C_{12}$ $H_{20}$ $O_{10}$ + 2 $H_2$ O. Sugar is frequently found dissolved in the cell-sap, being the product of a transformation of the preceding substances. Sugar is known by its having a sweet taste, and by being colored rose-red by sulphuric acid in the presence of protein substances. When a solution of sugar is mixed with caustic potassa, and a solution of copper is added, the liquid assumes a deep blue color. By the application of heat, and even at the ordinary temperature, the copper-oxyde is reduced, and red protoxide of copper is separated. If sugar dissolved in water come in contact with protein substances, or other agents of fermentation, it is soon decomposed. Various acids are originated, both by the spontaneous and chemical decomposition of sugar.

Besides the kinds of sugar already enumerated, various substances occur in plants, which, on account of their sweet taste, would seem to be entitled to be classed amongst the sugars. Many chemists and physiological botanists have indeed so classified them, but in their other properties they so resemble the extractive or neutral substances that we deem it more proper to treat of them in that connection.

PECTIN.— *Vegetable Gelatine.*—This substance, in the moist state, forms a colorless, tasteless jelly. When deprived of its water by expression, it becomes opaque, forming a fibrous mass, in which may be traced the outlines of organic structures. When thoroughly dried, it may be reduced to powder. Placed in water, it swells up gradually, and on standing for a time, the whole becomes reduced to a clear, transparent liquid, which, on the addition of alcohol, salts, or sugar, becomes gelatinous. Boiled in dilute sulphuric acid, it yields *grape-sugar*. When boiled with a solution of caustic alkali, a clear liquid is produced—pectinate of alkali—out of which, by the addition of acids, pectic acid is separated in the form of gelatine.

Pectin occurs in all plants, but is found in the greatest quantity in juicy fruits and fleshy roots. It is the product of a regressive metamorphosis, and constitutes the first condition in the series of changes which cellulose undergoes when it is being reconverted by the vegetable activity into primary nutritive subtances.

GUM.—($C_{12} H_{20} O_{10}$.)—Gum is found in all plants, and in

CONSTITUENTS OF PLANTS. 49

every part of their structure, dissolved in the cell-sap. It exudes from the ruptured bark of various trees, particularly of the genus *Acacia, Prunus,* and *Amygdalus*. In the recent state, gum is liquid, but soon hardens on exposure to the air. When sugar in solution comes in contact with protein substances, fermentation ensues, and among the products of the decomposition which takes place, we find a peculiar gum.

Gum occurs in the form of an amorphous, transparent, or semi-transparent, bright and brittle mass, without smell, or taste. It is soluble in water, forming a frothy, sticky liquid, but is insoluble in alcohol and ether. It combines with bases, and when boiled with dilute sulphuric acid yields *grape-sugar*. Gum in solution is precipitated by basic acetate of lead, (3 Pb O + $\overline{A}$,) and gives as a product, 2 Pb O + $C_{12} H_{20} O_{10}$.

When a solution of gum is mixed with a solution of caustic potassa, and sulphate of copper is added, a blue precipitate, composed of gum and oxide of copper is thrown down. This precipitate is not changed in color by boiling. Gum, like the preceding substance, is also a product of vegetable metamorphosis.

MUCILAGE.—($C_{12} H_{20} O_{10}$.)—This substance, like the preceding, is also common to all plants, and occurs in a similar manner. It exudes from the ruptured bark of many trees and plants, either pure or mixed with gum. It forms the covering of many seeds, and is a constituent of many roots. Mucilage forms an amorphous, semi-transparent, tough mass, without smell or taste. When mixed with water it swells considerably, forming a sort of jelly, its particles being suspended in a partial state of solution. It is insoluble in alcohol and ether. When boiled with dilute sulphuric acid, it is first converted into *gum*, and finally, by continued boiling, into *grape-sugar*. Mucilage combines with bases.

This substance was for a long time considered to be identical with gum, but the essential characteristics of the two are so dissimilar that we marvel much at their being confounded. Gum dissolves readily in water, while mucilage, which is composed of *arabin, bassorin, cerasin, &c.,* simply swells up, forming a gelatinous mass. Mucilage, like the preceding article, is only a peculiar condition of metamorphosed vegetable material.

There are several articles of therapeutic value recommended in medicine, which, from their properties of swelling up in water, and forming gelatinous masses, would seem to belong to this class. Among these are the *Calendulin* from the *Calendula Officinalis*, and the *Trilliin*, from the *Trillium Pendulum*. But their other properties, for instance their solubility in alcohol, would seem to entitle them to a distinct classification. In view of these distinctive characteristics, therefore, we propose for this class the name of *Muciresins*.

*Viscin* is a glutinous substance obtained from the berries of the *Viscum Album* or *Misletoe*. It is not a particular chemical compound, being only a product of the decomposition of the cellulose contained in the outer-cells of the misletoe seeds.

Many inorganic elements enter the structure of plants, forming therein various chemical compounds. Amongst the more important of these substances, we might mention various earthy matters, alkalies and alkaline earths, metallic oxides, particularly those of iron, alumina, manganese, &c. In general we find the alkalies and alkaline earths combined with inorganic or organic acids, forming salts; while the proper earths and metallic substances are mostly combined with inorganic matters, particularly with the coloring matters. The alkaline salts are found dissolved in the cell-sap; while the salts of the alkaline earths are suspended in the form of crystals in the cell-sap, and amongst the cell secretions. If either the cell-sap or the cell secretions are extracted from the plant, these crystals are also extracted, still retaining their original form. If this extracted substance contains resins, resinoids, or oleoresins, they are precipitated on the addition of water, being insoluble in that menstruum, and mechanically carry down at the same time these earthy crystals. This mechanical combination is so strong, that a great complication of chemical processes would be rendered necessary to overcome this admixture. In fact it would be impossible to effect a separation without injuring, and oftentimes destroying the properties of the various proximate principles. Ignorance of these facts has given rise to unjust and malevolent charges of impurity and adulteration of concentrated remedies. But malice and chemical ignorance are alike unavailing, and aspiring tyros may "hide their diminished heads" in the presence of the stern array of facts we now adduce. Honest and capable criticism is the great conservator of medical science; but the puerile vaporings of the mercenary and incompetent sometimes cast a blighting *incubus* over the motives and labors of those who are honestly striving to advance the interests of true science. As we have stated in the preceding chapter, the coloring matters of plants are often possessed of valuable remedial properties. They are often combined, as above stated, with earthy and metallic substances, from which they cannot be separated without effecting their decomposition. Hence it will be seen that the retention of the coloring principles in the concentrated remedies is based upon sound philosophical and chemical authority; and instead of militating against their value, confirms them in the possession of the aggregate medicinal value of the plant.

Among the bases common to plants, potassa, soda, lime and magnesia predominate. Of acids, the sulphuric, phosphoric, carbonic, tartaric, vinic, oxalic, and malic occur most frequently. The organic acids are in general combined with the above mentioned bases, forming acidulous salts. Silica is found in the cell-walls of nearly all plants, and oftentimes in considerable quantity. These inorganic constituents of plants are not accidental admixtures, but act as important agents in the processes of vegetable activity. Their particular influence seems to partake more of a *catalytic* than a *chemical* character.

FIXED OILS.—These substances occur in plants suspended in the cell-sap, in the form of minute drops or globules of variable size. They are more abundant in seeds, but are found in lesser quantities in a great number of plants. The fixed oils are soluble to a greater or lesser extent in alcohol and ether, but are insoluble in water. They are saponified by a solution of caustic alkalies, and are then soluble in water. The strong light-refractive power of the fixed-oil drops, and the fact of their disappearing under the action of caustic alkalies, enables us to detect them, even in the smallest quantity, by means of the microscope.

Fixed Oils are a mixture of *margarin* and *elain*, the former being a compound of *margarinic acid*, ($C_{34}$ $H_{20}$ $O_{3}$) and the latter of *elaic acid*, ($C_{41}$ $H_{80}$ $O_{4}$) with *glycerine*, ($C_3$ $H_4$ $O_2$) which answers the purposes of a base. They are often colorless, but in general possess a distinctive color, in consequence of their holding in solution certain absorbed coloring matters. In many plants they are substituted for starch and its metamorphosed conditions. They are also liable to similar transformations, of which we have an example in the germinating of an oily seed, in which instance the fixed oil affords the proper materials for the formation of cell-substance. By similar transformations a great number of products are originated, but of many of them we know but very little. In general they have an acidulous, or electro-negative reaction. The fixed oils are very dissimilar in their composition, although they all conform in containing carbon, hydrogen, and oxygen. Only a small part of them can be considered as simple organic oxydes, while by far the greater number of them are salt-like compounds possessing different degrees of fusibility. The number of these salt-like compounds is very large, the more common and greater part of them being employed for technical and economical purposes, a small number only possessing medicinal value. Each one of these compounds consists of a peculiar fat, having an acidulous reaction, (arising from the presence of *sebacic acid*,) neutralised by an indifferent organic oxyde, which oxyde cannot be separated without being altered in its composition. When treated with strong inorganic bases, these compounds are decomposed, the sebacic acid unites with the base, forming a *sebate*, and the indifferent organic oxyde is set free. While this decomposition is taking place, the organic oxyde absorbs the elements of water, and appears in an altered condition. Every fixed oil possessing a peculiar medicinal value, yields, by this process, a peculiar acid, which acid, however, does not entirely conform in its therapeutic reaction with the compound from which it is derived. The acids thus artificially produced, occur also in plants, being originated by the metamorphosis of vegetable material.

A part of the fixed oils, when exposed to the air, absorb oxygen, discharge carbonic acid, and are converted into a resinous substance. Another portion of them, when similarly exposed, simply dry down to a soft, greasy mass.

WAX.—Of this substance we have many varieties. They do not, however, form a distinct class of substances, but belong properly to the class of fixed oils. Like the latter, they are salt-like compounds, consisting of a fatty acid (*cerain*) united with an indifferent organic oxyde. Wax is never found in a liquid form, but always of a solid consistence, somewhat soft and unctious to the touch at a common temperature, but hard and brittle when exposed to the cold. It is but very slightly soluble in cold alcohol, somewhat more so in hot alcohol, but readily dissolves in the fixed and volatile oils, &c. Wax occurs in many plants, forming in many instances a thin granular coating upon the epidermis. It also is found as a coating upon the berries of certain plants, as the Bayberry or Wax Myrtle, (*Myrica Cerifera*.) Wax possesses but feeble, if any, therapeutic power. Its use is mostly confined to the preparation of plasters, ointments, and other external appliances.

VOLATILE OILS.—These substances are of very frequent occurrence in plants, but are mostly confined to certain organs or groups of cells. Where they exist in but small quantity, they are generally dissolved in the cell-sap; but they are often found occupying the entire cell, as well as the spaces between the cells.

Volatile oils, like the preceding substances, are gradually developed by progressive vegetable activity. The greater number of them are liquid at a common temperature. A few are solid, but very readily fusible. They are for the most part colorless. A small portion are colored, and may be either yellow, green, blue, red, or brown. They all possess a penetrating odor, and a warm, pungent taste. When brought in contact with paper, a transparent stain is produced, which disappears upon the application of heat. Many liquid volatile oils hold solid volatile oils in solution. When allowed to stand undisturbed for a period of time, or at a low temperature, the latter are separated in a crystalline form and precipitated. This precipitate is termed *stearoptene*.

Volatile oils are but very slightly soluble in water, the solution simply acquiring in a slight degree the odor and taste of the oil. They are readily soluble in alcohol, ether, and liquid fats. Only those volatile oils that contain oxygen are soluble in dilute alcohol. The greater the proportion of oxygen they contain, the greater their solubility. The boiling point of volatile oils is variable. The greater part of them, however, boil at 320° F. When distilled alone they are partially decomposed. Most of the volatile oils contain oxygen; a large number, however, are destitute of oxygen. When exposed to the air, they absorb oxygen, give off carbonic acid gas, and are finally converted into resins. Some volatile oils become acidulous by the absorption of oxygen, and gradually separate crystals of a peculiar acid. Alkaloids convert them, in the presence of air, into resins, with which they enter into combination forming resinates. A few of these oils contain sulphur. Volatile oils are not alone produced during the period of organic activity, but are frequently originated by fermentation, or the reaction of nitrogenous or oxydising substances upon indifferent vegetable materials. Many living plants contain no volatile oil; but as soon as they cease to grow, and are subjected to fermentation, volatile oils are originated. We have an illustration in the volatile oil of bitter almonds. This oil does not exist ready formed in the almond, but is originated by the reactions which take place while it is undergoing decomposition. By powdering the kernels coarsely, mixing them with water, and allowing them to stand for 24 hours, a peculiar fermentation ensues. Two products are originated by this fermentation, viz, *volatile oil*, and *hydrocyanic acid*. This is brought about by the reaction of *emulsin*, a peculiar nitrogenous substance, upon *amygdalin*, whereby the latter is decomposed, and the two above-named substances are produced.

The CAMPHORS are nothing more nor less than solid volatile oils, (stearoptene.)

Many plants owe their employment in medicine to the volatile oils they afford.

RESIN—Resins are peculiar proximate principles, possessing different degrees of solidity. They are mostly hard, brittle, and pulverulent; sometimes soft, and, when they exist mixed with volatile oils, semi-liquid. The solid resins are non-conductors of electricity; but, when subjected to friction, become electro-negative. A small number only are crystallisable. The specific gravity of the larger number of them is greater than that of water, ranging from .9 to 1.2. All resins are fusible, some being decomposed, others not, but none can be volatilised without undergoing decomposition. They are inflammable, and burn with a bright, but smoky flame. Solid resins undergo no alteration when exposed to the air; but soft and semi-liquid resins gradually harden, by reason of the volatile oil being converted by degrees into resin.

The origination of resins out of volatile oils is effected in various ways. For instance, a certain amount of oxygen is absorbed from the atmosphere, and an equivalent amount of hydrogen is displaced, resulting in a degree of oxydation. In the second place, a certain quantity of oxygen is absorbed without displacing the elements of water, and in the third place, by the absorption of a larger quantity of oxygen, with or without displacing hydrogen, but forming and discharging carbonic acid gas. This last reaction results in a higher degree of oxydation.

All resins contain carbon, hydrogen and oxygen. They are insoluble in water, but soluble in alcohol, ether, volatile and fatty oils. They do not unite with acids, but, on the contrary, many of them have an acidulous reaction, which is shown by their changing blue litmus paper to red. They combine readily with bases, forming *resinates*, and are freely soluble in a solution of caustic or carbonic alkali. Acidulous resins, when dissolved in alcohol, are not precipitated by the addition of ammonia, but those possessing no acid character are thrown down. Resins sometimes occur in the cells of plants, but in general exist in the form of secretions outside of the cell. They are truly nothing more nor less than oxydised volatile oils, and are often artificially produced by the reaction of acids, or a higher temperature, upon organic substances. Resins form an important class of remedial agents.

BALSAMS are simply a mixture of resin and volatile oil.

OLEO-RESINS.—The substances designated by this appellation do not form a separate class, possessing distinct chemical characteristics, but are simply a mixture of resin, wax, and fixed oil. These compounds are mostly found in the leaves and stems of plants, and are generally of a greenish color. The wax and fatty oil, admixed with the resin, seldom have any particular medicinal value. The therapeutic properties reside chiefly in the admixed resins.

GUM-RESINS.—Gum-resins are likewise a mixture of different substances, which are found circulating through certain of the cell-vessels, particularly the bast-cells, of various plants. When these cells are ruptured, the gum-resin exudes out, and, on coming in contact with the atmosphere, hardens, forming a brown or yellowish gray mass. When it first exudes from the plant it resembles the milk-sap in appearance, and is of a white or yellowish cast.

Gum-resins are only partly soluble in water, with which they simply form an emulsion; neither are they wholly soluble in alcohol. In general they are composed of a mixture of different resins, gum, mucilage, volatile oil, and in some instances, alkaloids.

RESINOIDS.—Like the preceding, these are also a mixture of different vegetable constituents. They are formed by a combination of several resins possessing different degrees of electro-negative reaction, and of dissimilar chemical properties. A portion of these resins, when separated, are readily and wholly soluble in a solution of caustic ammonia, and a saturated solution may be boiled for a short time without separating the resin. When the solution is evaporated, a compound of ammonia and resin remain behind, in which the resin largely predominates. This portion of the resins have an acidulous reaction, and are strongly electro-negative.

Another portion of these resinous constituents are also soluble in a solution of caustic ammonia at the ordinary temperature; but when the solution is boiled for a quarter of an hour, the resin is separated free of ammonia. Although less electro-negative than the preceding class, they are precipitated from their alcoholic solution by the addition of a solution of acetate of copper. They are soluble, by the aid of heat, in a solution of carbonate of soda, carbonic acid being expelled during the process. Their alcoholic solution reddens blue litmus paper.

A third class of these resins are neither soluble in a solution of caustic ammonia, nor in a boiling solution of carbonate of soda, but are readily soluble in a solution of caustic potassa or caustic soda. They are not precipitated from their alcoholic solution by the acetate of copper; but give a precipitate when treated with an alcoholic solution of the acetate of lead. Their alcoholic solution, when hot, reddens blue litmus paper. This class of resins are feebly electro-negative.

A fourth class of these resins are insoluble even in a solution of caustic potassa or soda, but may sometimes be dissolved in a saturated alkaline solution of some other resin, from which they are again precipitated on the addition of more of the alkalie. These resins have no acidulous or electro-negative reaction, and form a distinct class, for which we propose the term *indifferent* or *neutral resins*. Resinoids are insoluble in water, but are completely soluble in alcohol. They form a common constituent of plants, and are produced chiefly in those bast-cells having a milk-sap circulation. None are colorless, neither uniform in color, varying from light to dark yellow, red, brown, or green. Plants are richest in resinoids at the period when vegetable activity is arrested by the approach of winter. By the reaction of strong acids, resinoids are decomposed and converted into tanneous substances. The greater number of them combine with tannic acid and form compounds insoluble in alcohol. All resinoids possess, without exception, valuable remedial properties.

CAOUTCHOUC.—This substance occurs only in bast-cells having a milk-sap circulation, and appears in the form of small globules suspended in the milk-sap, giving to it an emulsion-like appearance. It is extensively employed for technical purposes, but cannot be considered as a therapeutic principle, seldom or never being used internally.

COLORING MATTERS.—These substances are diffused throughout the entire structure of plants, and give to them their characteristic colors. They differ very much in their chemical properties, many of them being soluble in water, and bearing a strong resemblance to the *neutral* or *extractive substances;* while others are insoluble in water, but soluble in alcohol, and bear a strong similarity to the *resins.* Another portion are soluble in ether, and conform in their general properties with *wax.* Many of them combine with acids and bases, whereby their original color is greatly modified, and their properties changed. The larger number of them combine with metallic oxydes, and form insoluble compounds. A few of the coloring matters contain nitrogen.

The most common of the coloring matters is *chlorophyl.* It is found in all plants, and gives to them their green color. Chlorophyl contains nitrogen. It generally appears in the form of minute grains; but these grains are not wholly composed of chlorophyl, being a mixture of chlorophyl and colorless protein substances, the latter largely predominating in quantity. Chlorophyl is produced directly from protein substances by the action of sun-light. It is insoluble in water, but soluble in alcohol, ether, hydrochloric acid, and solutions of alkalies. During the fall months it undergoes a series of changes, being converted into a red coloring matter termed *erythrophyl,* or into yellow colored substances called *xantho-*

*l.* The peculiar processes by which these changes are produced are not understood. Coloring matters, in general, are found dissolved in the cell-sap, or suspended in it in small globules. We know of but a few instances in which the cell-walls themselves are colored, while at the same time the cell-sap is colorless; but we often find cells entirely filled up with coloring matters, particularly with those possessing tanneous properties. We also often find coloring matters existing in the form of secretions outside of the cells, and which, in general, partake of a resinous character. When the cell-sap is of a red color, it indicates the presence of free acid. Blue cell-sap indicates the presence of free alkalie. The coloring matters increase in amount, in dried plants, for a series of years. Many of these coloring principles possess valuable therapeutic properties, while others are wholly inert and worthless as medicine.

EXTRACTIVE SUBSTANCES OR NEUTRALS.—These terms are applied to a great number of substances which may be extracted from plants by means of water or alcohol, either cold or hot, and which possess very different physical and chemical properties. They are called neutral because they have neither basic nor acid properties, and possess neither positive nor negative electricity. Many of the neutral substances are crystallisable, others are amorphous; some are colorless, but in general they possess a distinctive color, varying according to the source from whence they are derived. But few of the neutrals are tasteless. They may be either sweet, bitter, astringent, or sharp and caustic. When a solution of these neutral substances is exposed for a time to the action of the atmosphere, they are materially altered in their composition and properties. Particularly is this the case when a watery solution is evaporated. During the process of evaporation oxygen is absorbed, carbonic acid is formed and escapes in the form of gas, and a dark brown substance but slightly soluble in water is separated. This extractive sediment is termed *apotheme*. This apothemean substance first appears upon the surface of the liquid in the form of a brown colored pellicle, and finally precipitates in the form of a powder. This substance continues to form so long as the evaporation is continued, or so long as any of the neutral principle remains. Neutral substances may be instantly converted into apotheme by the action of *chlorine*. The conversion of neutral substances into apotheme is brought about by the absorption of oxygen from the atmosphere, which combines with a portion of the carbon of the neutral, forming carbonic acid, which is expelled, while at the same time a part of the hydrogen and oxygen of the neutral unite and form water. For this reason the apotheme appears richer in carbon than the neutral from which it is derived. Apotheme is slightly soluble in cold alcohol, more so in hot, and easily and readily dissolved in a solution of carbonic or caustic alkalie, out of which solutions it is again precipitated by the addition of acids, with a portion of which it combines. Apotheme bears a striking resemblance to *ulmic, humic,* and *japonic acids,* according to the neutral from which it is derived.

The neutrals agree in their general character in one respect only; and that is, in being very easily and readily decomposed, the slightest influence of other substances, when brought in contact, being sufficient to produce their complete decomposition and destruction, and entirely change their chemical and therapeutic properties. In other respects they are very dissimilar, having no general character in common.

A solution of some of the neutral substances is readily absorbed by charcoal, either animal or vegetable; while others are absorbed only after long continued boiling. Others are precipitated from their solution by the addition of a solution of the tri-basic acetate of lead, ($3\ Pb\ O + \overline{A}$). Many of the neutrals are insoluble in absolute alcohol. Many of them are remarkable for their hygroscopic properties, absorbing moisture from the atmosphere very rapidly.

Neutral principles occur in all plants, and form a large and important class of proximate medicinal principles. Many plants are entirely dependent upon the possession of this principle for remedial value and to its presence owe their employment in medicine.

*Humus* and its metamorphosed products are not constituents of the living cells, and have no therapeutic value. They are products of the decomposition and putrefaction of solid organic matters, and are important only as belonging to the class of nutritive substances which the plant absorbs from the soil.

Having now completed our brief history of the principal constituents of vegetable activity, we propose to give the *rationale* of the reactions whereby the proximate medicinal principles are decomposed, and their chemical composition and therapeutic properties either modified or entirely changed. In order that the reader may more readily trace the application of our exposition, we will first briefly recapitulate the main facts in relation thereto, set forth in the preceding pages.

We have seen that acids are originated during the period of organic activity, and also by the decomposition of some of the constituents of the plant after the cessation of that period. A greater number of the vegetable acids pertain to the nutritive constituents, and are originated by the conversion of amylum and oils into cell-substance. These acids are reconvertible into amylum and oils. Tannin substances, it has been shown, are not products of organic growth, but are formed by a putrefactive conversion of cellulose. With protein substances they form insoluble compounds, and by the action of water and oxygen are converted into huminoid substances.

Alkaloids are the products of living cells only, never increasing in amount in the dead plant, yet liable to form combinations whereby their medicinal value is suppressed. Thus, they combine with tannic acid, and when the latter is present in sufficient quantity, form bi-tannates, which are insoluble except in stronger acids.

Cellulose, xylogen, amylum, dextrine, pectin, gum, and mucilage all agree in yielding grape-sugar. Cellulose is converted by the living cells into wood and cork substance, and by retrogressive chemical action into starch. Protein substances undergo spontaneous decomposition, and transfer this property to substances of a more enduring composition, thus greatly promoting putrefaction and decomposition.

Starch is converted by the processes of vegetable activity into dextrine and grape-sugar, out of which products cellulose is formed by the action of protoplasma. By the same activity cellulose is reconverted into starch. By the reaction of albuminous substances, starch is converted, first into dextrine, then into grape-sugar, and finally into acetic and lactic acids.

Sugars are liable to both spontaneous and chemical decompositions, by which various acids are originated, and which in turn react upon the therapeutic principles.

Gum combines with bases, and is sometimes originated by the reaction of protein substances upon a solution of sugar.

Mucilages also combine with bases, and sometimes with resins, forming a class of substances to which we have given the name of muci-resins.

Fixed oils are substituted for starch in some plants, and are liable to similar decompositions, and give origin to a variety of acids.

Volatile oils are converted by the action of the atmosphere into resins. By the absorption of oxygen some of them become acidulous, and deposit crystals of a peculiar acid. Alkalies convert them, when exposed to the air, into resins, with which they combine and form resinates. Many volatile oils do not exist in the living plant, but are originated by the reaction of nitrogenous or oxydising substances upon neutral vegetable materials, during the process of decomposition.

Resins are frequently acidulous, and combine with bases, forming resinates. They are often artificially produced by the reaction of acids, or of a higher temperature, upon organic matters, particularly volatile oils.

Balsams, oleo-resins, and resinoids are compound substances, containing two or more of the previously described proximate principles. Resinoids are converted by the action of strong acids into tanneous substances. The greater number of them combine with tannic acids, and form compounds insoluble in alcohol.

Neutral principles form a large class of remedial substances, constituting the entire medicinal value of many plants. They are remarkably susceptible to disorganising influences, and are readily and rapidly decomposed. By the evaporation of their watery solution they are converted into a peculiar substance termed *apotheme*. They are remarkable for the avidity with which they absorb water, hence should be carefully excluded from the air. They are the principles usually afforded by aqueous preparations, and the first to undergo decomposition in pharmaceutical preparations. We shall notice them again in speaking of extracts.

Those compound vegetable substances which are least complicated in their structure, that is, which contain the smallest number of elements, as well as the smallest number of equivalents, or atoms of the elements of which they are composed, are the most constant and enduring in their character, and longest resist decomposition. In proportion as the number of elements, or the number of the atoms or equivalents of the component elements is increased, do organic compounds manifest a disposition to undergo transformations, and to resolve into more simple forms. The presence of water and oxygen is sufficient, at the ordinary temperature, to institute and promote those peculiar decompositions which are variously termed fermentation, putrefaction, moldering, and rotting. Every substance which will absorb water and repel oxygen, or which will combine directly with organic compounds, rendering them more permanent in their composition, will prevent or retard decomposition.

The processes by which the different kinds of decomposition above-named are effected, are various and dissimilar, accordingly as the substance is exposed to a free access of air, or is immersed in water, or buried in the soil. The most simple form of decomposition is that which organic substances containing no nitrogen undergo, when exposed to the action of the atmosphere. The *rationale* of the process is as follows—

*Oxygen* is absorbed, combines with the *hydrogen* of the organic substance, and forms *water;* while the *carbon* and *oxygen* of the substance unite and form *carbonic acid* which is dissipated in the form of gas. This species of decomposition is simply a process of *oxydation.* Woody substances particularly, undergo this variety of decomposition when exposed to the necessary conditions. It is this species of decomposition which trees, cut in the early part of the season, when the stem is succulent with sap, so rapidly undergo. It is for this reason that trees cut at this period are unfit for the purposes of timber, as we have already stated in the preceding chapter. Nor are they of much value as firewood, for by this spontaneous decomposition, a greater portion of their *carbon* and *hydrogen* is, so to speak, *burned up;* that is, consumed in the formation of *carbonic acid gas,* and in this form dissipated.

Medicinal roots, barks, etc., gathered at this season, are subject to the same species of decomposition, and speedily become inert and worthless.

Another species of decomposition to which organic substances containing no nitrogen are subject, takes place when those substances are brought in contact with water, and partially excluded from atmospheric air. In this instance not only is *oxygen* absorbed, but also the *elements of water*, which are taken up in considerable quantities, *carbonic acid gas* is expelled, and the result is a compound possessed of very different chemical and physical properties. Woody fibre is peculiarly subject to this species of decomposition, and manifests the change by gradually losing its color, density, and becoming pulverulent. This phenomenon is frequently to be observed in the stems of old trees. In familiar language it is termed *powder-post*.

Not even the complete exclusion of atmospheric air will prevent non-nitrogenous organic substances from undergoing decomposition, provided *water* be present. Of this we have an example in the formation of bituminous and anthracite coals.

Organic acids, even when chemically pure, cannot be preserved in the form of a watery solution, without being decom-

CONSTITUENTS OF PLANTS.    65

posed. For example, the *oxalic acid*, ($C_2 O_3$,) the most simple in its composition of all the organic acids, is speedily decomposed when dissolved in water, no matter how effectually it is excluded from the air. In this instance the water is decomposed, and its elements uniting with those of the acid, various products are originated, as follows : a portion of the *oxygen* of the water combines with a portion of the *carbon* of the *acid*, and forms *carbonic acid;* while a portion of the *hydrogen* of the *water* combines with a portion of the *carbon* of the *acid* forming a *hydro-carbon* compound, (C. $H_2$,) etc. At the same time a peculiar *fungus* is generated, belonging to the lower order of cryptogamic plants, known in common language by the name of *mould*.

A solution of *tartaric* and other *acids* will undergo decomposition in a manner similar to that which we have just described.

Those organic substances which contain *nitrogen*, evince a more ready tendency to undergo decomposition than the preceding, provided the volume of *nitrogen* bears a due proportion to the rest of the elements; that is, if the number of its atoms be neither too great nor too small. By the addition of another element, the affinities of a more simple substance are increased, and these affinities being displayed at the same time, the consequent reactions become more complex. Vegetable matters containing *nitrogen*, absorb *oxygen* from the atmosphere, and decompose the water which may be present. The *hydrogen* which is set free by the decomposition of the water, combines with the *nitrogen* of the organic matter, and forms *ammonia*, ($N.H_3$); while the *oxygen* either unites with the *carbon* of the vegetable material and forms *carbonic acid* (C. $O_2$), or with *carbon* and a portion of the *hydrogen* of the water, forming some other organic acid, as, for instance, *lactic acid* ($C_6 H_5 O_5$ + H. O.) At the same time *carbonic acid* and another portion of the *hydrogen* are set free, and if the organic substance contains *sulphate* or *phosphate salts*, they will be found to yield traces of *sulphuretted* or *phosphuretted hydrogen*, (H. S.) or (P. $H_3$). When nitrogenous substances are immersed in water.

5

they not only undergo spontaneous decomposition, but also decompose a portion of the water, forming *carbonic acid*, ($CO_2$) and *carburetted hydrogen* ($CH_2$), which escape in the form of gas; while certain *huminoid* products, poor in *oxygen* and *hydrogen*, and mixed with *salts of oxyde of ammonia*, remain. The unstable character of nitrogenous substances renders them great instigators and promoters of organic decompositions; and being capable of transfering this disposition to the substances with which they may come in contact, they frequently induce decomposition in organic compounds which would otherwise resist disintegrating influences for a great length of time. In consequence of this transferred property, the constituents of one substance frequently unite with those of another, when brought in contact. This blending of the constituents of the two substances does not always occur, however; but the disposition of the one substance to undergo decomposition is attended with a consequent activity of its particles, which are set in motion, and these communicating their influence to the particles of the passive substance, overcome the indolence of their chemical affinity, and induce certain changes or transformations in its composition. The first species of decomposition is called putrefaction; the second, fermentation.

When nitrogenous and non-nitrogenous vegetable substances are commingled, and undergo putrefaction together, their constituents reciprocally react upon each other, effecting their mutual decomposition, and their elements reuniting in different numbers and proportions, various new products are formed; as, for instance, *ammonia, lactic acid, carbonic acid, carburetted hydrogen, butyric acid, mannite, gum, mucilage, etc.*, and various offensive *gases*, fumes of which are emitted during the progress of the mutual reactions. This species of decomposition is sometimes termed *mucinous fermentation*.

In the instance above cited, if the oxygen cannot be derived from the atmosphere, by reason of its exclusion, it is obtained by the reduction of the admixed substances; as, for instance, *water* and *sulphate salts*, the latter being conveited into *sulphurets*. Thus, if *sulphate* of *iron* be present, its oxygen is absorbed, and it is converted into *sulphuret* of *iron*. The *hydrogen* which is set free by the decomposition of the water, frequently combines, at the moment of its liberation, with fragmentary portions of the admixed substances, forming products rich in hydrogen. This species of putrefaction takes place when the leaves of the *indigofera*, and other plants yielding the blue indigo of commerce, are immersed in water. A kind of mucous fermentation ensues, water is decomposed, carbonic acid, ammonia, and hydrogen gases are evolved, and the particles of coloring matter, which were blue in the leaf, are held in solution, colorless. This loss of color is effected by the combination of hydrogen, set free by the decomposition of water, with the blue indigo. The formula of blue indigo is $N. C_{16} . H_5 O_2$. That of white indigo is $C_{16} H_5 .N. O_2 + H$. It will be seen that the latter contains an excess of hydrogen. By the absorption of oxygen the blue color is restored.

The term *fermentation* is generally applied to the decomposition which sugar undergoes when exposed to the action of nitrogenous substances; but upon referring to the preceding definitions and illustrations, and taking into consideration the great number of organic substances generated by the processes previously described, it will be seen that the greater number of organic destructive processes belong to this species of decomposition.

A great many organic substances may be produced by fermentation artificially excited.

The fermentative decomposition of sugar is excited and promoted by the introduction into its solution of a peculiar cellular fungus, termed yeast. This substance consists of small, cell-like globules, of which it is estimated that a cubic inch contains eleven hundred and fifty-two millions. The cell-walls of these globules are isomeric in composition with starch, while the contents consist of a peculiar protein substance which very readily and speedily undergoes decomposition. The formation of this fungus is effected by the action of the oxygen of the atmosphere upon protein substances held in solution. During the process of fermentation, the protein contents of these fungus cells are decomposed, and acetate of ammonia and other products are formed, leaving the cells empty and exhausted. As soon as fermentation begins in the cells of the yeast, it is transferred to the particles of the sugar, which is converted thereby into carbonic acid and alcohol. When fermentation is once generated in contact with the atmosphere, it does not cease when the fermenting substance is excluded from the air, even if immersed under water. The necessary oxygen is obtained by the reduction of water, and of sugar, while at the same time new products, rich in hydrogen, are formed. If the substances undergoing fermentation are neutral, among the products will be found fusel-oil; if the fermenting liquid be acidulous, ether-like compounds are formed. By a temperature above 80° of Farenheit, fermentation is changed into putrefaction, and sugar is then converted into mannite, gum, lactic acid, and butyric acid.

It is by a similar process of fermentation that *amygdalin*, when acted upon by *emulsin* in the presence of water, is decomposed, giving rise to the formation of *volatile oil, hydrocyanic acid, sugar, carbonic acid,* and *carburetted hydrogen*.

*Amygdalin* belongs to the class of crystallisable neutral substances. Its formula is $C_{40}H_{27}.N.O_{22}$. It is soluble in water, out of which it crystallises in large colorless prisms, containing six equivalents of water. Amygdalin is derived from the kernels of the bitter almond, peach, cherry, prune and other fruit stones, and from the bark of the wild cherry, choke cherry, etc.

*Emulsin* is a peculiar nitrogenous constituent of both the sweet and bitter almond, in the former of which it exists independent of the presence of amygdalin. It is also found in the kernels of other fruits, and in many plants. It is soluble in water. The formula is $N.C_{16}H_{12}O_2$. In the cotyledon of the almond and other fruits, and in those other plants yielding these two principles, the amygdalin and emulsin exist in separate and distinct cells, hence, in that condition, cannot react upon each other; but when these substances are bruised or reduced to powder, and mixed with water, fermentation and decomposition immediately ensue. Not only are the amygdalin and emulsin decomposed, but also a portion of the water, giving rise to the products above named. We shall have occasion to refer to this peculiar fermentative decomposition when treating of the properties and employment of Prunin and Cerasein.

A similar fermentation ensues when *sinapisine* is submitted to the action of *myrosyne*, in the presence of water. Decomposition takes place, and volatile oil, sulpho-hydrocyanic acid, and other products are originated,

We have now completed our brief history of the different varieties of decomposition to which vegetable substances are subject, and propose, in another chapter, to make a practical application of the preceding facts while discussing the subject of officinal preparations. Our great aim is to awaken the attention of the profession to the best methods of preparing remedials, be they either simple or compound, so that we may secure their *full*, and what is of quite as much importance, their *definite* value. If we can show that the ordinary pharmaceutical preparations, such as are now recognized by the term *officinal*, are defective and not prepared in accordance with the requirements of science, we will have made a beginning. But if we can go further, and point out the manner in which these imperfections may be rectified, we feel that we shall be **truly advancing the interests** of positive medical **science.**

# CHAPTER III.

## CONCENTRATED MEDICINES.

Officinal Preparations—Infusions—Decoctions—Extracts, etc.—their liability to Decomposition—of variable strength—frequently inert, etc —Concentrated Medicines proper—their advantages—uniform and definite in strength—not liable to change, etc.

From the earliest times many disadvantages have been recognised in the employment of crude medicines, and many and various processes have been devised, whereby to bring their remedial properties into a more definite and convenient form. We have shown some of the disadvantages arising from the employment of crude organic medicines in the first chapter, and, as there promised, will now endeavor to demonstrate the correctness of the objections there named.

We now propose to critically examine the various methods of preparing organic remedies for the use of the physician, and to apply the tests of organic chemistry to the preparations named at the head of this chapter. We shall then be enabled to see how far the labors of the pharmaceutist have tended to accomplish the desired object. And first we will examine—

INFUSIONS.—These constitute the most simple form in which vegetable remedial substances are prepared for exhibition. A part, or the whole of a plant is bruised and put into some convenient vessel, boiling water is added, and the whole is allowed to stand for a time in a warm place. The hot water softens and swells up the tissues of the plant, and extracts a portion of those principles which are soluble in water, both medicinal and nutritive. Thus far the process seems well enough; but let us look a little closer at the process, and examine it in all its aspects. In the first chapter we have demonstrated that one great objection to the employment of crude organic remedies, depends upon the fact that they are extremely variable as regards the quantitive product of active proximate principles. Not only does the actual amount of medicinal constituents vary infinitely, but frequently are the specimens entirely inert. Will this discrepancy be equalised by the preparation of the substance in infusion? By no means whatever. Not only will the therapeutic deficiencies of the plant go unremedied, but absolutely be rendered more uncertain in consequence of the presence of a considerable amount of other active principles *insoluble in water*, such as resins, resinoids, oils, etc., which will not only be retained by the plant, but will also cause the retention of a considerable amount of such portions as would be otherwise soluble. In this way the full value of such of the medicinal power as would otherwise be yielded to the water, is withheld. As plants are richer at some seasons than at others in those principles which are insoluble in water, it follows that the gathering of the plant at different seasons, will exercise a great influence in modifying the character of the infusion.

In the second place, when certain insoluble active principles are present in the plant employed, they cannot be rejected without seriously impairing the value of the preparation. When a number of therapeutic properties are attributed to a plant, we naturally infer that those properties respectively reside in separate and distinct proximate principles, and do not look for them to be concentrated in one single isolated principle. Hence the disappointment frequently experienced by the physician, by reason of overlooking the question of plurality and solubility of the active medicinal constituents of plants.

Thirdly, the water not only extracts a portion of the medicinal constituents of the plant, but also a greater part of the nutritive or non-medicinal substances, such as grape-sugar, gum, mucilage, dextrine, pectin, various acids, protein substances, tanneous principles, etc., which of themselves undergo spontaneous decomposition, and the accompanying fermentation involves the certain destruction of whatever medicinal constituents may be present. Thus are infusions speedily rendered worthless, the time required varying from a few hours to a few days. By referring to the preceding chapter, the reader will be enabled to comprehend the nature of the reactions liable to ensue when the above named substances are mingled in solution. He there will find the individual characteristics of the different constituents defined. Thus if the plant be prized on account of its yielding tannin, and at the same time protein substances are afforded to the solution, they will combine with the tannin, forming an insoluble compound. Or if no protein substance be present, the tannin is shortly converted, by the action of water and oxygen, into humus. Or on the other hand, if these conditions are not present, and a neutral principle is held in solution independent of the presence of other substances, it is converted, if evaporation take place, into apotheme. If the plant yield a soluble alkaloid, and at the same time tannic acid, they will combine, forming, if the quantity of tannic acid be considerable, a bi-tannate, which is insoluble in every menstruum except a stronger acid. Other changes are liable to take place, which the reader may easily determine by consulting the preceding chapter. We have particularised a few, in order to account for the disappointment which no doubt many physicians have experienced in the employment of medicinal plants when prepared in this form. In consequence of the want of a knowledge of these facts, many really valuable plants have been condemned as worthless, or at least as of uncertain value. The natural defects of the plant, together with the unscientific method of its preparation, have created much division of opinion, and brought many excellent remedies into disrepute.

The chemical properties of the water employed in making

infusions, exercise a greatly modifying influence. Thus, if the water contain earthy salts, as, for instance, carbonate of lime, it will precipitate a great proportion of the medicinal constituents, and render the infusion comparatively worthless.

DECOCTIONS.—When a plant is boiled in water for a time, the solution so formed is termed a decoction. Such preparations generally contain more of the soluble constituents of the plant, particularly the nutritive, a portion of which become insoluble when the decoction cools, and are precipitated. These precipitates, in falling, mechanically carry down a considerable portion of the medicinal matters held in solution, and thus materially diminish the value of the preparation. By reason of their containing more constituents, their chemical affinities are increased, and their tendency to decomposition augmented in proportion. As stated in the first chapter, such preparations speedily run into the putrefactive fermentation, particularly in warm weather. If the plant yield tannic acid, it will combine with protein substances, forming insoluble compounds, and these, as above stated, will mechanically carry down much of the medicinal matter present. If the plant happen to yield a soluble alkaloid, it will combine with the tannic acid, and thus be rendered inert. It will be remembered that the age of the plant, dating from the period at which it was gathered, will make a great difference in regard to the amount of tannic acid present, as that constituent is formed after the death of the cells, by a putrefactive conversion of the cellulose. Hence the amount of tannic acid increases with age. We have, then, to contend with much uncertainty when a remedy is so prepared. First, the uncertain amount of medicinal power residing in the crude material; second, the liability of such of the medicinal constituents as may be extracted, to be precipitated from their solution; third, the rapid decompositions which take place when a number of vegetable constituents are mingled together in solution. Thus, if the plant yield starch, together with albuminous matters, and which are almost universally present, it will undergo decomposition, if the decoction be allowed to stand for a time, being first converted into dextrin, next into grape-sugar, and finally into acetic and lactic acids. If grape-sugar be afforded, it will also undergo decomposition, giving rise to the formation of various acids. These, in turn, will react upon other of the constituents present. If tannic acid be extracted, and does not combine with any other substance, it will be converted, after a time, by the action of the water and atmosphere, into humus. Thus are certain and complicated reactions involved, and the character of the preparation rendered uncertain and inert. The presence of any nutritive principle whatever, is antagonistic to the integrity of every pharmaceutical preparation. Complete isolation of the therapeutic constituents is the only safeguard.

EXTRACTS.—Of these preparations we have several varieties, termed respectively, aqueous, alcoholic, hydro-alcoholic, inspissated, and fluid. No department of pharmacy more needs a thorough reformation than this. While we are far from impugning the motives of those who manufacture these preparations for the use of the profession, believing that they have honestly and faithfully endeavored to effect the best results their knowledge of organic chemistry would permit, we nevertheless desire to call the attention of the profession to the obvious defects that pertain to such preparations, and invite their serious attention to a consideration of the facts we are about to present in relation thereto.

AQUEOUS EXTRACTS.—When an infusion or decoction is evaporated to a syrupy or honey-like consistency, the residue is known by the general term of *extract*. It will be remembered that the usual and almost sole medicinal constituent yielded to water is a *neutral principle*. It will also be remembered that we stated, when treating of the chemical properties of neutral principles that when their watery solution is exposed to a free access of air, and evaporated, they undergo a material alteration in their composition. This is precisely what occurs in the preparation of watery extracts. The continual change of air to which the surface of the evaporating liquid is exposed, gives rise to the formation of a peculiar substance much resembling *humus*, to which the name of *apotheme* has been given. This substance, as it forms, is precipitated, and in common with the concentrated nutritive substances which may have been afforded by the plant, forms the ordinary aqueous extract. If the evaporating liquid is exposed to a strong heat, the neutral principle is completely decomposed, and the extract rendered entirely worthless for all medicinal purposes.

This change does not take place to so great an extent when the extracts are prepared *in vacuo;* but even then they are rendered none the less liable to the spontaneous decompositions which afterwards ensue, as sufficient water will always be present to institute and promote a destructive metamorphosis. Frequently are the plants employed in making extracts nearly or quite destitute of any proximate active principles whatever, in which case we have nothing for our pains but a worthless mass of starch, grape-sugar, protein substances, gum, pectin, etc. Thus lean we upon a broken reed perhaps, in the time of our greatest need. Or perhaps the active principles that give medicinal value to the plant are insoluble in water, and again is such a preparation obviously worthless. Even admitting, for the sake of argument, that the watery extract may have secured the neutral principle unchanged, yet a very short time will suffice to render the preparation valueless. This result will arise from the fact of their admixture with those non-medicinal constituents which so readily and rapidly undergo decomposition, and which, as stated in the preceding chapter, communicate their disposition to substances of a more enduring texture. As neutral principles are the first to be affected by such decomposition, it follows that those preparations depending upon the presence of this principle for therapeutic value, will soonest be rendered worthless. All *extracts* become entirely inert in one year from the time of their preparation. Extracts of *narcotic* plants are generally worthless after the expiration of six months. Some extracts are entirely decomposed at the end of three weeks. Extracts that should be kept in hermetically closed vessels, are frequently put into earthen pots with loosely fitting covers, and thus exposed to the destructive ravages of air and moisture.

ALCOHOLIC EXTRACTS.—These are prepared by digesting the crude materials in alcohol, of various per centages, until the medicinal constituents are dissolved out, and the solution so formed is reduced by distillation or evaporation to the proper consistency. Extracts so prepared are preferable to the preceding, inasmuch as they contain a lesser proportion of the non-medicinal constituents of the plant, provided the alcohol employed in their preparation be not too much diluted. Yet even these will contain grape-sugar, tanneous substances, various acids, and water, quite sufficient to cause them to undergo decomposition in a very short time. It is also a common practice amongst extract makers, to boil the materials, after they have been exhausted with alcohol, in water, so long as they will yield any soluble matters, and to add this watery product to the alcoholic solution. In this way the *quantity* of extract is increased, but the *quality* is impaired, as the added constituents consist of gum, starch, grape-sugar, pectin, dextrin, and other non-medicinal matters, all active agents in promoting fermentation and decomposition. Even when excluded effectually from the action of the atmosphere, extracts are not proof against decomposition, as water is always present in sufficient quantity to stimulate the chemical affinities of the non-nitrogenous constituents, and when once the fermentative or putrefactive processes are commenced, their influence, as previously explained, is communicated to the more resisting constituents. The more complex such a preparation may be, the greater the number of its affinities, consequently liable to a greater number and variety of chemical reactions.

*Hydro-alcoholic extracts* are similar to the above, the only difference being that the plant is exhausted, or, we might more properly say, digested in dilute alcohol. Hence less of the constituents requiring strong alcohol to dissolve them are obtained, while the non-medicinal nutritive substances are chiefly extracted. Their defects are therefore self-evident.

INSPISSATED EXTRACTS.—This name is given to preparations made by reducing the expressed juice of the fresh plant to the proper consistency. The plant is bruised and subjected to pressure, and after all the juice that is possible is obtained in this manner, hot water is added to the plant, and the pressure again applied, and so on *ad finem* until all the properties are supposed to be extracted. The solutions so obtained are mixed, and exposed to a heat above 150° F., in order to coagulate the protein substances, filtered, and evaporated to the proper consistence. Extracts so prepared are similar to the alcoholic extracts, except that if the plant from which they are obtained contain an alkaloid principle, which generally occurs in a crystalline form, the extract will not possess it, unless the alkaloid be soluble in water, which, however, is seldom the case. Pressure will not extract the crystalline alkaloid principles of plants.

FLUID EXTRACTS.—These are old preparations with new titles. They are variously prepared, and are nothing more than infusions, decoctions, or tinctures, reduced to a semi-fluid or syrupy consistency. In some cases the plants are treated with water, the solution evaporated, and a quantity of alcohol added. At other times, the evaporated solution is mixed with syrup or molasses, and the required consistency thus obtained. Sometimes alcohol is employed as the menstruum, which, however, is generally evaporated off, and sugar and water substituted. The vapor of alcohol, or water, or both, is employed by others, but in either case no definite result is arrived at, so far as regards the medicinal strength of the preparations. They possess no advantages over other extracts, being neither definite nor uniform in therapeutic power. As a general thing, they contain very little of the active medicinal constituents of the plants from which they are derived, and frequently none at all. In this statement, we are supported by the experience of eminent and scientific professional men, their judgment in the matter being rendered only after carefully conducted and extensive trials. In a paper read before the New York Academy of Medicine, by a distinguished member of that body, an impartial history is given of numerous clinical and chemical analyses, and these preparations there proven to be variable, uncertain, and frequently inert. The reader has but to transfer the application of the foregoing facts to these preparations, and thereby save us the necessity of a recapitulation. They are open to all the objections and accidents of other extracts, differing only in degree.

TINCTURES.—The ordinary tinctures of organic medicinal substances are fully as indefinite in remedial power, as any of the preparations we have been describing. As in all other instances, *physical* considerations alone are the criterion for their preparation. A given amount, by *weight*, of some crude substance is directed to be added to a given quantity of alcohol, by *measure*, and this completes the process, except, in some instances, digesting and filtering. The alcohol employed is seldom of uniform strength, and frequently the alcohol employed in preparing tinctures from the same plant is of variable per centage. We have amply demonstrated the fact that all crude organic materials are never uniform in their yield of active principles, and this alone would prove the character of ordinary tinctures unreliable. But other considerations may be appropriately cited. If the alcohol is not of sufficient strength, a great proportion of the active principles requiring strong alcohol to dissolve them, such as resins, resinoids, oils, etc., will not be extracted, while at the same time a larger amount of the nutritive constituents are taken up, such as grape-sugar, etc., and the tincture thus rendered more susceptible of decomposition. Not even tinctures are proof against change and decomposition, although they suffer to a less extent than the previously described preparations. Tinctures which, when newly made, have neither alkaline nor acid reactions, become, after standing for a length of time, acidulous, as is proved by their power of reddening blue litmus paper. When tinctures are allowed to stand undisturbed for from three to six months, be they ever so securely stoppered, they will give a brownish colored precipitate. This precipitate belongs, in consideration of its chemical properties, to the class of humoid products, which proves abundantly that the medicinal constituents have been undergoing decomposition. Alcohol will not prevent tannic acid from undergoing decomposition, neither will it prevent the catalytic influence of one constituent over another. Alcohol, when diluted, is an excitant of decomposition, as may be seen by observing the process pursued in some parts of the country for making vinegar.

SYRUPS.—These are simply fluid extracts mixed with cane-sugar. The extracts may be either aqueous, alcoholic, or hydro-alcoholic. The usual proportion of sugar employed is two parts to one of extract. It is generally conceived, and so stated, that this proportion of sugar will prevent decomposition of both the medicinal and nutritive substances. This, however, is an error. Sugar simply *retards*, but will not *prevent* decomposition. That it will not prevent decomposition, we have many familiar examples in domestic economy. For instance, when certain unripe fruits are preserved in pure syrup of sugar, they gradually ripen, and in course of time become matured in flavor and other characteristic properties. This proves that a material metamorphosis has taken place in their constituents, retarded perhaps, but not prevented by the presence of the sugar. Walnuts gathered while yet unripe, and before any traces whatever of oil can be detected in them, and preserved in sugar, will undergo a progressive change in their constitution, and the characteristic oil of the nut will be developed. Thus we see that even sugar will not hold the chemic forces in abeyance. But we need not go beyond the dispensatories to prove that sugar will not prevent decomposition. It is there admitted. Directions are also given, that if a syrup ferment, it be re-heated, and again allowed to cool. It is also stated that " a syrup thus recovered is less apt to undergo subsequent change, on account of the fermenting principles having been decreased or consumed." What are these "fermenting principles" so "decreased or consumed?" It must be borne in mind that syrups depend chiefly for their medicinal value upon the presence of neutral principles. These principles, it must also be remembered, are extremely susceptible to decomposing influences. No neutral principle can preserve its integrity in the presence of fermentation once excited. They are the first to undergo change in all fermenting solutions. Nor is much time required to effect their total destruction. These are a part of the "fermenting principles" which are "decreased or consumed" when syrups manifest a disposition to decompose. We would caution practitioners not to risk the lives of their patients, nor their own reputations, by using syrups which have fermented and been re-heated, as the "fermenting principles decreased or consumed" by the process constitute, in nine cases out of ten, the sole remedial properties of the preparations.

Sugar does possess, in a degree, the power of preventing direct oxydation, by reason of its property of absorbing oxygen. It also may retard decomposition for a time by absorbing water. But this will not prevent the reactions of the constituents upon each other. These reactions partake, in many instances, of a catalytic nature. Cane-sugar, by the absorption of two equivalents of water, is converted into grape-sugar, and its power to prevent oxydation or decomposition is thereby materially lessened. This conversion proves that certain reactions have been going forward. But while cane-sugar has the property of absorbing water, it must be borne in mind that some of the neutral principles also possess this property in a preeminent degree. To demonstrate this fact, we have but to expose a quantity of cane-sugar and an equal amount of certain of the neutral principles to the action of the atmosphere, when it will be found that, while the sugar becomes dried, actually losing a portion of its moisture, or remains unchanged, the neutral principles will absorb water and harden. For the purpose of experiment, Leptandrin, Cypripedin, Populin, etc., may be employed, each of which contains a neutral principle possessing hygroscopic properties. This phenomenon as readily takes place in syrups and other preparations, and affords an illustration of elective affinity. It will be seen, therefore, that sugar would be inefficient in preserving such constituents from undergoing change. This property of the neutral principles exercises a most important bearing upon the history of all crude organic remedies, and will explain the variable and uncertain character of many remedial agents.

We have now enumerated the principal defects of the foregoing classes of pharmaceutical preparations. We are aware that the processes we have mentioned are frequently varied, and that other solvent menstrua are frequently employed, such as ether, wine, etc., but the main features of the case are not thereby altered. We do not entertain the idea for a moment that aught we have said will bring these preparations into entire disuse. But we hope, nevertheless, that practitioners will give their serious consideration to the facts we have advanced. We have given the chemical proofs, step by step, and we doubt not that the experience of all observing physicians will confirm the truth of our exposition. Although the facts in the case have long been apparent, no explanation has hitherto been given which might serve to reconcile the various opinions relative to the remedial value of many plants.

Every practitioner of medicine is well aware that there are times when it is difficult to decide upon the precise remedy indicated. Of this fact we have an illustration in every consultation held over a case of disease. When, at last, combined judgment has decided upon the proper remedy, greatly is the perplexity increased, if it be of indefinite therapeutic power. Be it either above or below the common standard of medicinal strength, mischief will equally happen. If too strong, the reaction may prove fatal. If it be inferior or inert, valuable time will be lost, and the chances of recovery lessened, if not destroyed. How often does the reputation of the skillful practitioner suffer, by reason of the dispensation of such defective agents as we have enumerated. His diagnosis may have been perfect, his judgment correct, his prescription appropriate; yet, in consequence of the dispensation of extracts, tinctures, syrups, etc., prepared from inferior, perhaps inert materials, his patient fails of receiving benefit, his judgment is impugned, his prescription condemned, and his reputation injured. The positive character of any and every remedial agent is a consideration of the greatest importance to every practicing physician. It is a consideration that directly involves the question of success. No conscientious physician would risk his patient's health and life by the employment of remedies of doubtful power. Even were the practice of medicine made a purely commercial transaction, yet would the hope of further patronage be based upon the power to cure. In either view of the case, then, the positive character of remedial agents is a question of great moment. Health, life, success, reputation, hang in the balance. We feel that we cannot be too strenuous upon this point, and are certain that all right-thinking physicians will coincide with us in the opinion that all remedial agents should be as *positive* and *definite* in their character as human skill can make them. The substances of the inorganic materia medica have been defined with great precision and care, and why should not those of the organic? In former years the attention of chemists has been more especially devoted to a consideration of the substances of the mineral kingdom, consequently greater progress has been made in that department of chemical science. It is only of late that the subject of organic chemistry has received that attention to which its importance justly entitles it, and its true bearing on the interests of practice fully appreciated. Within a few years, however, the attention of the profession has been directed to the development of this branch of the collateral medical sciences, and already are good results flowing in upon us from this fountain of scientific industry. Though yet but in its initial flow, still may we with reason anticipate that the patient industry of coming years will expand this little rivulet into a broad and noble stream, upon whose placid bosom the physician may with safety launch his therapeutic bark, and guided by the tide of truth, and impelled by the motive winds of duty and philanthrophy, "carry healing to all the nations."

We now come to a consideration of

CONCENTRATED MEDICINES PROPER.—In the history of all the more important medicinal plants, we find a record of the various attempts which have been made to ascertain upon what particular constituents they depended for therapeutic value. But one prevailing error has rendered the majority of these attempts abortive. This error consisted in conceiving that multiple therapeutic powers could reside in one single constituent. Thus, an oil, a resin, a resinoid, or an alkaloid was supposed to embody the entire therapeutic constitution of the plant. This conception not holding good in practice, it followed, in many instances, that the attempt to ascertain this peculiar constituent was abandoned, and the plant continued to be employed in the ordinary manner; while in other instances the vending of these isolated, fragmentary resin, resinoid, or alkaloid preparations, represented as being *the* active constituents of plants, has brought many excellent remedies into disrepute. This is not to be wondered at when we consider that, in procuring these fractional constituents, the more valuable proximate medicinal principles were rejected as worthless, and out of some three or four active principles, some one resin or resinoid only preserved. It is true that a number of isolated alkaloid principles are esteemed of great value in medicine. Usually such principles are limited in the number of their therapeutic properties, possessing in general but one or two well defined powers. Thus in morphia we have the principal narcotic power of opium. But no one will say that morphia is equivalent to opium. Morphia is esteemed especially as an anodyne and soporific Opium is considered narcotic, sedative, stimulant, astringent, anti-spasmodic, febrifuge, diaphoretic, etc. Quinia is but one of a number of active principles belonging to the Peruvian bark. It represents the anti-periodic tonic power of the bark. So with many other similar preparations that we might mention. Resins are generally possessed of but limited and feeble powers. Resinoids are remarkable for possessing a greater number of distinct therapeutic powers. The reason of this we have explained in the preceding chapter, under the head of resinoids. We have there shown that they are compound substances, composed of a number of different resins. We have enumerated and described four varieties, but some resinoids are more complex still. Each one of these resins has a different chemical character, behaves differently towards reagents, and possesses individual electric properties. It would be philosophical, therefore, to suppose each resin to be possessed of different therapeutic properties, which is absolutely the case. The resinoid principle of Podophyllin has been separated into five resins, and we have reason to suspect a greater complexity in its constitution. This will account in a measure for the great number of physiological results which that remedy is capable of producing. So we might run on through the whole organic materia medica, eliciting facts all tending to prove that the diverse therapeutic properties of plants reside, not in *one*, but in *many* principles. We have shown how, in the preparation of extracts, etc., the neutral principles of plants are altered in their composition or completely destroyed. We have also shown that they constitute an important class of proximate active principles. We claim to have been the first to recognise their true remedial value, and the first to have established their identity as a class of distinct proximate principles. We were also the first to record their physical and chemical characteristics. We likewise claim to have established the existence of a new class of proximate active principles, to which we have given the name of *muci-resins*. In view of all these facts, it must be evident to the reader that, in order to secure the *full* value of a medicinal plant, these various proximate principles should be isolated from all extraneous combination, and then recombined. This is precisely what has been done in the preparation of the concentrated medicines treated of in this volume. Every plant has been carefully and repeatedly analysed, and both its physical and therapeutical constitutions definitely ascertained. In the prosecution of these investigations, much patient labor has been bestowed, and the elevation of pharmaceutical sci ence the ultimate object. The results have been gratifying to those engaged in conducting the investigations, and, we trust, beneficial to the interests of positive medical science. We are now enabled to define the number and character of the proximate active principles of plants with greater accuracy than has hitherto been attained. In making this statement we design to cast no reflections upon the motives and labors of others, cheerfully recognising and admitting their claims to whatever of advancement they have made, simply reserving to ourselves the credit of having detected and explained many of the errors and defects of organic chemistry as at present conducted, and, consequently, to have made greater progress and improvement in this department of pharmacy than any other organic chemists, by their productions, have yet secured. There may be those engaged in this field of scientific labor who will yet outstrip us in our efforts to perfect the character of organic concentrated medicines. If so, we shall be amongst the first to recognize and rejoice at their success, and to gratefully acknowledge their superior claims in having advanced the interests of progressive medical science. As yet, however, we believe that the concentrated medicines prepared at the laboratory of B. Keith & Co. are superior to all others yet offered to the profession. Our reasons for this opinion we will now endeavor to state. In the first place, they are not fragmentary preparations, composed of a single resin, resinoid or alkaloid principle, but combine *all* the active medicinal principles of the plants from which they are severally derived. The only exception to this rule is when a plant yields an oil, in which case it will not be present in the powdered preparations. When this is the case, the fact is stated. In the concentrated tinctures the oil, if there be any, is included.

To illustrate the advantages of having all the active constituents of a plant combined, we will take the article of *Podophyllin*. By reference to the article treating of the properties and employment of this agent, it will be seen that it is composed of **three active principles**, viz, a resinoid, alkaloid, and neutral. Thus combined, the action of this agent is modified, and its operation rendered comparatively mild, while at the same time its therapeutic powers are increased. All other specimens of *Podophyllin* we have ever seen, consisted of the resinoid principle alone. This principle, like all other resinoids, is insoluble in the stomach, and soluble only in the enteric secretions. It also possesses a degree of escharotic power, which, in certain inflamed conditions of the glandular surface of the intestines, renders its employment objectionable, in consequence of its peculiar irritating properties. This action arises chiefly in consequence of the derangement of the functions of certain of the glands, whereby their secreting and absorbing powers are diminished or suppressed. If the secreting power be suppressed, the resinoid, not meeting with the proper solvents, will remain undissolved, and act as a mechanical irritant. Even admitting the existence of activity on the part of the absorbent vessels, the resinoid must yet be in a state of solution before it can be absorbed. If, on the other hand, the secreting vessels are active and the absorbent functions suspended, the resinoid, although it pass into a state of solution, will be retained and expend its influence locally, and thus add to the existing irritation. With the *Podophyllin* combining the three active principles of the plant, this local influence will be found to be essentially modified. The neutral and alkaloid principles, which exist in a state of combination, are soluble in the stomach, and are generally directly absorbed, producing a specific effect upon the glandular system before the resinoid has yet had time to act. In this way the diathesis of the system is changed and corrected, and the requisite conditions for the further action of the remedy are secured. It is in consequence of the soluble character of the neutral and alkaloid principles that *Podophyllin* frequently manifest so speedy a control over the functions of the system. Many symptoms are allayed, and decided sanative results produced, ere the resinoid principle has had time to pass the pylorus and be reduced to a state of solution. Another reason why the resinoid principle is sometimes so much of an irritant, is the fact of the presence of a minute quantity of a very acrid oil, which adheres to the resinoid, and which appears to be a protoplastic resinoid principle not yet matured. It is found only when the plant is gathered at an improper season, and while the development of the proximate constituents is yet incomplete.

The *Leptandrin* is another remedy combining a number of important active principles. The *Leptandrin* of which we shall have to treat contains *three* more proximate medicinal principles than the *Leptandrin* of other manufacturers, viz.: a resin, neutral, and alkaloid. Hence its range of application and therapeutic powers are proportionately increased. So with all the concentrated medicines of which we shall have to treat, with the exception of *Geranin* and *Myricin*, they being the two only remedies with which we are acquainted, of other manufacture, that contain more than one principle, and these consist, in each instance, of a resinoid and tannic acid. They are, therefore, the two only remedies that embody the total active value of the plants from which they are obtained.

We think we may justly claim, therefore, that the concentrated remedies described in this work are superior and more nearly complete than any yet offered to the profession. We claim that they are the *concentrated equivalents* of the plants from which they are severally derived, uniform in strength, definite and positive in therapeutic power, and will preserve their properties unchanged for an unlimited period of time. Their several principles are isolated singly, deprived of all foreign admixture, and then recombined in the same numbers and proportions as they existed in the plant, unchanged in composition, and entirely free from the presence of any of those non-medicinal constituents which we have shown are instrumental in effecting the decomposition of ordinary pharmaceutical preparations. Does any one doubt that these results can be accomplished? Does any one doubt the existence of Morphine, Quinine, Emetine, Jalapin, etc.? Are they not well defined, positive medical agents, uniform in therapeutic power, and capable of being preserved for an indefinite period of time? But, says the reader, these are simple alkaloid or resin principles. True, but if it is possible to isolate *one* single principle, is it not possible to isolate a *number* of single principles residing in the same plant? And if *one* is capable of being defined in chemical and therapeutic properties, is there any reason why *all* should not be? But there is no need of argument to prove that which is self-evident. The existence of these various principles so isolated constitutes the best evidence of the fact.

It is not consistent with the character of this work to give a detailed history of the various chemical processes involved in the preparation of these medicines. Such an exposition belongs properly to a more elaborate work on general materia medica. We are not writing a text book for chemists, but are endeavoring to embody those more important facts which daily concern the physician in the practice of his profession. In years past we have sadly felt the need of such information as we now have the pleasure of submitting to the profession, and we doubt not that our humble efforts will meet with a welcome response from all well-wishers of the art of healing. We have given the physiological and chemical history of each constituent under its appropriate head, and shall proceed directly to an exposition of the therapeutic properties and physiological effects of the combined active constituents. It is this portion of our subject that more nearly concerns the practitioner, whose province it is to administer medicines and not to make them. The manipulations of the laboratory come within the province of the chemist, whose duty it is to provide the physician with the means wherewith to execute the requirements of his profession. We have given a plain and truthful history of the active constituents of plants, and every physician is supposed to possess a sufficiency of chemical knowledge to enable him to test the correctness of our statements. Therefore he may easily satisfy himself as to the chemical character of the several preparations. But this will tell him nothing of their clinical value. The question with him is, are they *reliable* as remedial agents. This question is one that requires individual observation at the bedside in order to effect its most satisfactory solution.

So far as the writer is concerned, he bases his reputation as a practitioner, and as an author, upon the *positive* character of these preparations. Upwards of twenty years experience in collecting, curing, and preparing plants for medicinal use, and fourteen years experience in the clinical employment of organic remedies, both in their crude and concentrated forms, has given him a familiarity with the physiological effects produced by vegetable substances upon the human organism which enables him to pronounce the concentrated preparations, when all the active principles are combined, *fully equivalent to*, and *more reliable* than the plants from which they are severally derived, when prepared in any other form. Their curative action in disease is entirely analogous, and attended with greater certainty. Amongst the many advantages arising from the employment of organic remedies in this form, we esteem their *promptitude of action* a matter of the greatest importance. Being divested of all extraneous combination, they are purely medicinal; and as such, are prepared to act the moment they are taken into the system. Not so with crude remedies. When taken into the stomach, the latter require to undergo a digestive analysis, in order that the therapeutic constituents may be separated from their combination with those inert matters which are incapable of assimilation. In an enfeebled and disordered condition of the digestive apparatus, this taxation of its exhausted powers is a matter of serious moment, and its inability to perform this office will result in the withholding of the manifestation of any therapeutic power on the part of the substance so administered. This matter will either remain and act as a mechanical irritant, or pass off as useless ingesta. It is in this manner that we may account for the frequent failures of crude remedies, when administered in substance, in not producing their specific effects upon the system. Surely it is as reasonable to suppose that the stomach is as incapable at times of digesting crude barks, roots, etc., as it is of digesting bread, meat, etc. The skillful hygeist is scrupulously circumspect in appointing his patients diet, having reference to the digestive and assimilative power of his patients system;—then why should he not observe the same conditions and requirements in the appointment of his medicines. Even if the power to perform this office exist, considerable time must elapse before the medicinal constituents can be brought into a condition to admit of their appropriation. And even then, a considerable amount of inert matter frequently remains, imposing further taxation of the depurating organs to secure its removal. This of itself constitutes a serious objection to the employment of crude substances in cases of great debility. And again, the percolating through the alimentary canal of particles of woody matters and fermenting non-medicinal substances, frequently creates a serious disturbance of the nervous system, and with patients of a peculiarly susceptible organism, will often provoke a troublesome degree of spasmodic action. With children, this irritation will sometimes give rise to convulsions. We have seen the alvine discharges of patients who had dosed themselves, or been dosed with considerable quantities of powdered roots, etc., much resemble a mixture of saw-dust and water, when under the influence of a cathartic. The retention of such worthless material is quite as likely to provoke or prolong a febrile action as any other retained matter. In gastritis, enteritis, diarrhea, cholera morbus, dysentery, &c., the administration of vegetable remedies in substance, is bad practice, and can scarcely fail to aggravate the disease in every instance. Yet it is an error quite too common amongst some practitioners.

The remedies of which we shall presently treat, are free from these objections. Their composition and constitution is purely therapeutic, and they require neither an outlay of digestive action to prepare them for appropriation, nor the exercise of the functions of depuration for the expulsion of waste material. They therefore ensure a promptitude of action which can never attach to crude medicines, and thereby effect a saving of time which is frequently of the utmost importance. We would have every practitioner test the question of their reliability for himself, as we desire no one to be controlled by our judgement but respectfully ask that all will make the same impartial trial of their merits that we have done. Independence of action in this respect will give the practitioner a better conception of their remedial value than anything we may say concerning them. We simply give our own convictions, the fruits of a somewhat extensive clinical experience.

To sum up the advantages claimed for concentrated medicines combining the various active principles of the several plants, we pronounce them far superior to any yet offered to the profession, inasmuch as they are concentrated equivalents of the plants from which they are derived, entirely divested of all non-medicinal combination, positive in therapeutic power, uniform in strength, convenient of administration, and capable of preserving their properties unimpaired for a series of years. The only way in which the preservation of the active constituents of plants can be ensured, is to isolate them from all extraneous admixture, dry and reduce them to powder, and keep them in closely stopped bottles. It is not to be expected that a suspension of natural laws will take place in favor of the organic substances sooner than in favor of the inorganic. Light, heat, air, moisture, all conspire in executing the immutable laws of chemical trans-formations. The affinities of the atoms of matter are definite, fixed, and unchangeable. By the action of light, hydrocyanic acid, one of the most virulent of poisons, is decomposed and rendered inert. Iodine is volatile, and requires to be carefully excluded from the air. Chloride of zinc, various preparations of potassa, etc. absorb water and deliquesce. Hence certain precautions are necessary to the preservation of inorganic remedial substances. So with organic substances. By exposure to the air, certain volatile oils absorb oxygen, and are converted into resins. Those neutral principles possessing hygroscopic properties absorb water, harden, and become altered in their properties. When exposed to a strong light, some of the concentrated preparations will change in color, and as many of the coloring matters possess decidedly valuable remedial powers, they are thereby deteriorated.

Some objections have been held against the concentrated preparations on account of their not being decolorized. As we have shown in the second chapter, the retention of the coloring matters does not militate against the value of these preparations, but, on the contrary, confirms them in the possession of an additional therapeutic constituent. That coloring matters possess remedial properties, we have examples in the *cochineal insect*, and in *hematoxylin* derived from logwood, both of which are used medicinally. The coloring matters of plants are so intimately blended with the other active constituents that they cannot be separated without effecting their decomposition, and thus altering, and in many instances destroying their remedial properties. It will be seen, therefore, that the characteristic color of the various preparations, besides furnishing a distinguishing mark, denotes that no violence has been done in isolating their several principles.

In this connection we desire to speak of the *Concentrated Tinctures* prepared at the laboratory of B. Keith & Co. Their claim to superiority is based upon the same considerations as those of the powdered preparations, namely, their freedom from all inert admixture, positive character, uniformity of strength, and property of retaining their virtues for a great length of time unchanged. The process pursued in their preparation is the same as that observed in preparing the powders. Each active constituent is isolated singly, freed from all non-medicinal matters, and so on until the aggregate therapeutic principles are all obtained, which are then recombined and redissolved, in exact proportions, in alcohol of uniform per centage. This process ensures a certainty and uniformity in no other way attainable. Consequently the practitioner is as certain of the quantity of medicine he is administering, as he would be in exhibiting a definite solution of morphine, quinine, etc. Sufficient alcohol is employed to hold the various active principles in complete solution. When the plant contains a valuable oil or oleo-resinous principle, we deem this the better mode of preparation. With the more active plants, as the Veratrum Viride, Digitalis Purpurea, Hyoscyamus Niger, etc., this form of preparation is by many preferred. We have reason to believe that the tinctures operate more promptly under some circumstances than the powdered preparations, in consequence of their diffusible character. As a matter of convenience also, they offer some advantages, as the prescriber is saved the necessity of dividing them into separate doses. They also admit of a ready and convenient combination with each other, and in this way, as with the powders, their properties may be varied or increased. We will speak further of each under its appropriate head.

## ADMINISTRATION OF CONCENTRATED MEDICINES.

Success in the employment of remedial agents depends upon the observance of certain conditions. This is especially true of the organic medicines. Remedies ever so positive in therapeutic power, and uniform in strength, may yet fail of producing any specific effects upon the system. The first condition to which we would call attention, as being unfavorable to the action of the concentrated remedies, is the predominance of acidity in the stomach and bowels. We first called the attention of the profession to this subject, some two years since, through the medium of the medical journals, and we are glad to find that recent writers have adopted and reiterated our sentiments, although they have omitted, unintentionally, no doubt, to give us proper credit. Many practitioners, doubtless, have been disappointed in not realising anticipated results from the employment of concentrated remedies to which specific and positive therapeutic powers had been accredited, and from which, consequently, they were led to expect much. Frequently after a single trial, a good remedy has been condemned, simply because it failed to realise all that was expected of it, and because the true reason of the failure was misunderstood. In nearly all cases of disordered action there is a disposition on the part of the system to originate acid. In chlorosis, this condition incites the patient to seek after absorbents and alkalies, as magnesia, chalk, slate pencils, etc. In many cases the food, instead of being digested, undergoes a fermentative decomposition, and gives rise to the formation of various acids. Even the medicines administered, such as extracts, syrups, sweetened infusions, decoctions, and all preparations containing starch, sugar, etc., tend to aggravate the condition, by reason of their nutritive constituents undergoing fermentation. These acids very speedily decompose the resin, resinoid, and neutral principles, and hold the alkaloids in solution. The latter are not decomposed, but their action is suspended. In case an inorganic alkali is administered, being a stronger base, it robs them of their acid, and they are again set free. It is in this way that certain plants have gained the reputation of possessing cumulative properties. Repeated doses have been administered and failed to act in consequence of the acid present, which has combined with the alkaloid, when, by accident or spontaneous action, the acid has become neutralised, and the whole power of the accumulated remedy has been suddenly expended. Another reason we would assign for the apparent cumulative power of certain remedies, is the neglect of furnishing to the system the proper amount of fluid. In certain cases and stages of disease, when the fluids are greatly expended, this consideration is one of great importance. The physician's first duty, when called to prescribe in the advanced stages of typhoid and other fevers, is to supply the system with a proper quantity of diluent and demulcent drinks. When continued fever has occasioned a great expenditure of the serum of the blood, and the tongue, fauces, and mucous membranes of the stomach and bowels are dry and inflamed, it is bad practice to exhibit powders, or other solid substances in a scanty vehicle, as a little syrup, for by so doing the symptoms are aggravated and the object of the medicine defeated. The syrup will have but little effect in bringing about the necessary degree of solution, while on the other hand it will prove mischievous by reason of undergoing a fermentative decomposition. More acid is thereby formed, and the local irritation increased, while the medicine itself is liable to be decomposed and rendered inert.

Among the acids most destructive in their action upon the organic remedies, is the *lactic*. Podophyllin is not hindered in its operation by acetic acid, but the presence of a considerable quantity of lactic acid will almost entirely suppress its action. This will account for its failure in many instances in not producing its legitimate impression upon the system. We have known the operation of fifteen grains of the resinoid principle to be immediately checked and all further manifestations of therapeutic power arrested by the administration of sour milk. It is all important to the successful exhibition of organic remedies that undue acidity of the system be first neutralized. Attention to this necessity will save disappointment and loss of time, besides preventing many an excellent remedy from being unjustly condemned. Super-carbonate of soda is the most convenient antacid generally at hand, and may either precede or accompany the medicine. When the acidity is considerable, it is best to administer the soda half an hour before the medicine. From one half to one drachm is sometimes required. In other cases from five to ten grains will be sufficient. Common salt, chloride of sodium, will answer when soda cannot be obtained. But when a full dose of Podophyllin is administered, the too free use of salt during its operation will sometimes have a tendency to produce hyper-catharsis, while the remedy is in consequence liable to be unjustly blamed. We have repeatedly observed this fact.

The proper *combination* of concentrated remedies is a subject of much interest to the practitioner. Multiplicity of remedial agents is to be avoided as much as possible. We have observed, with regret, a fondness amongst physicians for numerical combinations. In the old dispensatories formulas are given for pharmaceutical compounds containing as high as sixty ingredients. The philosophy of the composition was, that where so many agents were combined, one, at least, would reach the case. The fact seems to be overlooked now, as then, that organic medicines are capable of and liable to mutual reactions, decompositions, and combinations. In this respect many of them are quite as susceptible as inorganic substances. Tannic acid will combine with vegetable alkaloids and render them insoluble. It will also almost entirely suppress the action of alteratives, particularly those designed to influence the liver. The practitioner may avoid the mistake of combining incompatable remedies by making it a point to treat diseases with simple substances, and to never add an adjunctive remedy unless a thorough knowledge of its influence over the remedy already administered, or the indications of the case render it justifiable. The true value of the concentrated remedies can never be estimated unless they are singly and thoroughly tested. One simple remedy will often answer a better purpose than half a dozen combined, although each one singly would be admissable and appropriate to the case. Many combinations may be judiciously formed, whereby the activity of a special therapeutic property may be augmented or modified, and by which the number and kind of remedial powers may be multiplied, instances of which we shall give in the following pages. Some writers have recommended the admixture of six and seven of the concentrated medicines, many of them incompatable and contra-indicated by the described features of the case. Such promiscuous combinations could have only been devised in the absence of practical knowledge, and a proclivity for plausible hypothesis. Brilliant theories in medicine are like the lightning's flashes; although they dazzle for a time, their explosion is followed by the thunders of discord, and intensified darkness. We were forcibly reminded of some formulas we have seen recommended for combining concentrated remedies by a prescription which recently came under our observation. It read as follows.:

R.
Comp. Fluid Extract of Sarsaparilla,
Simple Syrup,
Phytolacin,
Irisin,
Alcohol,

Con. Com. Stillingia Alterative,
Iodide of Potassium.

The Compound Fluid Extract Sarsaparilla contains five ingredients, viz.: *sarsaparilla, liquorice, sassafras, mezereon,* and *guaiacum.* The Concentrated Compound Stillingia Alterative contains seven ingredients, viz., *stillingia, corydalis, phytolacca, iris versicolor, xanthoxylum, chimaphila,* and *cardamon seeds.* Here are twelve ingredients besides the syrup, alcohol, and iodide of potassium. As to the *modus operandi* of such a combination we confess our entire ignorance. It may be a very scientific and eligible preparation, but we doubt whether its inventor could explain its precise therapeutic action, or how nature could ever succeed in unravelling the web of its composition. If all the therapeutic powers attributed to each single ingredient were to be displayed at the same time, we can imagine a very lively and complex excitement of the various functions of the system.

We would respectfully, yet earnestly, advise practitioners to observe simplicity as much as possible, assuring them that the best results will accrue from such a course. By closely observing the independent action of each remedy, he will be better enabled to judiciously effect proper combinations where occasion requires. Not only this, but he will also be able to distinguish the *remedy* from the *auxiliary*, a feature quite important in the treatment of disease.

Various suggestions have been made in regard to the manner of *administering* concentrated medicines. The trituration of the active principles with sugar is advocated by many. To this plan, however, we cannot yield our assent. We have already shown the impropriety of sweetened decoctions, syrups, &c., and can make no distinction between the latter and triturations with sugar. It is argued in favor of the employment of sugar, that it will prevent the local action of the medicine upon the stomach. This would seem to us to be an untenable position. In order to produce a local impression upon the stomach, the substances administered must be soluble in that organ. Will sugar prevent them from entering into solution?

If so, it will negative their action entirely, and their remedial influence will be lost. But such is not the case. Sugar will not *prevent* the local action of the remedy upon the stomach, but it will *diffuse* it. Again, the local action is one that is frequently desirable. All neutral principles are soluble in the stomach, and are absorbed directly by that organ. Sugar will not promote their solution, nor absorption. It only furnishes an additional, and, under the circumstances, an unnecessary constituent, requiring of itself to be digested and assimilated. If the stomach be competent, all is well. But if not, the sugar undergoes a fermentative decomposition, and gives rise to the production of acids which not only aggravate the existing disorder, but attack and decompose the accompanying active principles, and thereby destroy their power over the system.

Another argument in favor of the trituration of concentrated medicines with sugar is, that it enables them to become more readily absorbed and conveyed into the circulation. This we also deem an erroneous view. If the remedy and the sugar required the same solvents, no advantage would be gained, as the presence of the sugar would require more labor to be performed without any prospect of equivalent benefit. The sugar itself is not a solvent of the active principles, hence is of no utility in that respect. But as the constitution of the sugar and the concentrated medicines vary, it follows that different solvents are required, and that the dissolving, absorbing, and circulating of the active principles is an action quite independent of the presence of the sugar, which not only does not promote this action, but requires of itself to be similarly acted upon. Hence is a greater expenditure of digestive action occasioned to no purpose. We hold it a fixed and truthful principle in the practice of medicine, that the purer medicines are administered, and the less they are compounded with inert or nutritive matters, the more certain and satisfactory they are in their operation. Sugar is most objectionable in the treatment of many disorders of the digestive apparatus. We have succeeded in curing many cases of indigestion with the same remedies with which others have failed. They administered them in syrup, sweetened decoctions, etc., while we exhibited them in their purity, without sugar or other extraneous admixture, at the same time prohibiting the use of sugar and other fermentescible substances. Notwithstanding our objections to the use of sugar, we are in favor of triturating some of the concentrated medicines, with a view to their proper diffusion. We have devised and practiced a method of trituration which we now have the pleasure of submitting to the profession, and which we can assure them will answer a better purpose than any yet suggested. As most of the concentrated remedies are soluble in water, but few articles require triturating on that account. But with some of the more potent remedies, such as Veratrin, Podophyllin, Digitalin, Sanguinarin, etc., diffusion is desirable in consequence of the high degree of power attained by their concentration, and their more kindly operation when diffused over a larger nervous surface. Our plan is to triturate one agent with another. In this way is not only the desired object attained, but the activity of the remedy may be augmented or modified at the option of the practitioner. Our usual agent employed in trituration is the Asclepin. No remedy with which we are acquainted is so seldom contra-indicated as the Asclepin. In fact we do not know a single indication in which this remedy could be used amiss. By referring to the article on the employment of Asclepin, the reader may learn our reasons for so esteeming it. The Veratrin may be triturated with Asclepin in the proportion of one grain of the former to ten or more of the latter, at the option of the practitioner. The Asclepin will not only not counteract the Veratrin in any respect, but will enhance its diaphoretic property, an advantage instead of an objection, and an effect always desirable to be produced when Veratrin is indicated. The Podophyllin may be triturated in the same way, either with Asclepin or Caulophyllin, according as the diaphoretic or antispasmodic property may be desired. The Asclepin is nearly all soluble in water, and will render other of the concentrated remedies capable of being administered in that menstruum. So with the Caulophyllin. We shall treat more fully upon this subject in the second part of this volume, when detailing the employment of the concentrated medicines.

In the employment of the concentrated medicines combining the various active principles of the plant, combinations are not so frequently necessary as when single resin, resinoid, or alkaloid principles are used. Nearly all the remedies of which we shall have occasion to speak, possess several distinct and well marked therapeutic properties, hence are capable of fulfilling an equal number of indications. *Veratrin* is emetic, arterial sedative, diaphoretic, etc., and with it we may evacuate the stomach, reduce the force and frequency of the pulse, promote the cutaneous exhalations, abate febrile excitement, relieve local congestions, etc. *Populin* is diuretic, diaphoretic, febrifuge, tonic, etc. With it we may relieve and cure suppression and scalding of the urine, fevers, night sweats, indigestion, etc. Each remedy is already a natural combination in itself, and as such is generally better adapted to the necessities and assimilative powers of the system than any artificial combination.

In the constitution and arrangement of the active constituents of medicinal plants, we have a wonderful illustration of the wisdom and perfection of design of the Creator, in having so constituted and endowed the therapeutic atoms as to ensure perfect harmony of action when a number of distinct active principles are blended together. No clashing of adverse powers is observable when a single medicinal plant is employed. But when the assumptions of art have advised the indiscriminate commingling together of a great number of remedials, frequently is "confusion worse confounded," certainty reduced to uncertainty, and action and counter-action engaged in unprofitable warfare.

Following the discovery of vegetable alkaloids in 1816, the medical world was thrown into a fever of a decidedly alkaloid type. Physicians, chemists, druggists, apothecaries and the whole medical crew run r ad in the pursuit of what was supposed to be the *universus* of vegetable remedial powers. Creation was ransacked high and low, and simultaneous with the appearance of a purple stain upon a piece of reddened litmus paper, came the triumphant cry of *eureka!* But the ardor of the enthusiast was destined to be cooled by a succession of disappointments. Many alkaloids were found to be possessed of no particular medicinal value, while many plants were found destitute of any alkaloid principle whatever. And even where the alkaloid obtained was of value as a remedial agent, it failed to represent in full the therapeutic constitution of the plant from which it was derived. With a few exceptions, this class of agents have gone into disrepute. The medical profession have become satisfied that they do not fairly nor fully represent the remedial properties of the substances from which they are derived.

But notwithstanding the search after alkaloids failed of its purpose, much good has resulted from the investigations necessarily carried on. Other principles were brought to light, the existence of which was before unknown, or at least hypothetical. Resins and resinoids became the objects of the chemist's search, for still laboring under the *one* principle delusion, he sought to find in either of these *the* active principle of the plant. The alkaloid mania was not cured, but simply transferred. If the alcoholic tincture but threw down a precipitate when added to water, the long sought *desideratum* was thought to be obtained. No matter how much the water might hold in solution, or wash away, did but some insoluble matter remain, it was bottled up, vended as *the* active principle of the plant, and accredited with all the therapeutic powers pertaining thereto. Several preparations of this character are now before the profession, and we would caution them to critically examine all preparations purporting to be concentrated, and ascertain whether they actually combine the different active principles of the plant, or whether they are not, rather, fragmentary, resin or resinoid preparations only, and thus deceptive, being in truth *isolated* but *not concentrated.* And yet we have known those detached principles to receive the sanction of writers professing to high scientific culture, and assuming to be censors of the opinions and labors of others, and by them to be indorsed as *the* active principle, and as such pompously recorded in dispensatories, and other publications as among the *immense* discoveries of the nineteenth century. Now as these preparations are in each instance simply *an active principle, one of many*, how are we to relieve these authors of the dilemma in which they have placed themselves. If we attribute the error to a want of scientific knowledge, we shall most undoubtedly be visited with their direst indignation. If, then, we allow them, in charity, the credit of being perfect masters of the science of organic chemistry, how shall we relieve them of the seeming dishonesty which would lead them to palm off upon the profession these fractional resin, resinoid, and other defective preparations as being *the* active constituents of plants, instead of truthfully proclaiming them to be, what they really are, *isolations, one* of *several* active constituents, the rest having been lost, rejected, or their existence not known or suspected. We are inclined, however, to give them credit for honesty in one respect, and that is, in stating all they knew. But at the same time it would have been as well not to have been too positive of the dishonesty of others whose researches had fortunately resulted more successfully, and who had brought to light the several active constituents of the vegetable organism; and having made the discovery, and succeeded in isolating the various principles, adopted the rational idea of re-combining them as they existed in the plant. To those who were acquainted with single resin, resinoid, or alkaloid principles only, this combination of a number of principles was a new and startling idea, and many were inclined, honestly, we hope, to look upon it as an adulteration. But the better sense of the profession, as soon as informed of the true state of the case, generously yielded the credit due to those whose skill and penetration had secured the *real concentrated equivalents* of the various plants. Here we are willing to let the matter rest with the profession, having been drawn into make these remarks in consequence of some unjust aspersions having been cast upon the motives of those whom we believe to be honestly engaged in a good and important work. We acknowledge that the aspersions referred to came from sources which it would be derogatory to our self-respect to mention here, and we should not have mentioned the circumstance but to illustrate the liability of all discoverers to be maligned by the ignorant and viciously inclined.

To briefly recapitulate the most favorable conditions for the successful administration of concentrated medicines, we would advise that particular attention be paid to the neutralising of undue acidity, simplicity of combination, avoidance of the use of sugar and other fermentescible substances, and such general considerations in regard to diet, regimen, etc., as the circumstances of the case may render appropriate.

As a majority of the concentrated medicines are soluble or mixable in water, we would recommend that menstruum as being in general the best, as well as the most available. We are aware that many advocate the plan of rendering medicines as palatable to the patient as possible, entertaining the idea that their certainty and efficiency of action are governed in a measure by the likes and dislikes of the patient. With all due deference to the opinions of others, we would record our experience in favor of administering medicines in their purity as much as possible. Our experience goes to prove that much less medicine will usually be needed, while the specific influences of the remedy will in no wise be diminished. Podophyllin will ne'er fail of producing its usual effects in consequence of being disgusting to the palate. Hyoscyamin will alleviate pain, and induce a quiescent condition of the nervous system, despite the objections of the patient to its nauseous taste. We have never found a medicine to fail of its accustomed operation in consequence of its unpleasant flavor. We impress upon our patients the fact that we give medicines to cure disease, and not to please the palate; and we teach them to expect that any remedy possessing power to remove disease, must give some indication to the senses of its peculiar properties. We direct their minds to a consideration of the beneficial results to follow, and discourage all reference to its unpalatableness. The smallness of the dose when concentrated medicines are employed, renders disguise less frequently necessary. Our objections to foreign admixture have already been set forth in the preceding pages, hence there is no need to recapitulate them here. Pills may be formed in many instances, as a matter of convenience, to secure a more eligible form, and to overcome the objections held against the taste of the various remedies.

We shall now proceed to give a practical exposition of the therapeutic properties and clinical employment of such of the concentrated medicines as combine the different principles of the various plants. We wish it distinctly understood that our remarks apply only to such concentrated medicines as are prepared in accordance with the above conditions, that is, which combine the several active constituents of the plant. We do not profess to be able to give a history of all the indications which may be successfully fulfilled with these remedies, nor to enumerate all the combinations that may be judiciously and advantageously effected. We shall endeavor to give a truthful synopsis of the therapeutic properties characterising each remedy, relying upon the judgment of the practitioner to select such as are best adapted to the various necessities of the system.

The formulas we give are such as we employ in daily practice; and all reference to their curative action is a simple record of our own experience, except when expressly stated to the contrary. When our own experimental knowledge of a remedy is limited, we shall give the experience of those practicing physicians whose testimony may be received as reliable.

We are aware that disease is tempered by climatic and other influences, and that the treatment which proves successful in our own locality will require to be modified to meet the peculiarities of other sections. Yet the properties of the remedies will be the same in all climates, and the modifications required will be in regard to *combination, quantity, time, repetition* and *continuance.*

# PART II.

## Concentrated Medicines Proper.

Before entering upon the therapeutic and clinical history of the Concentrated Medicines Proper, we deem it due to the enterprise, energy, and industry of B. Keith, M.D., that proper credit should be here awarded him for his successful efforts in providing the profession with concentrated preparations of a definite, reliable, and uniform therapeutic character.

Of long experience in the clinical employment of crude organic remedies, his attention was early attracted to an investigation into the merits of so-called concentrated medicines. Upon testing these preparations in practice, he found a marked discrepancy between the therapeutic action of the "active principles," so-called, and the plants from which they were derived. To ascertain the cause of this discrepancy, and to provide the profession with *true concentrated equivalents* of the various medicinal plants, became the engrossing object of his scientific labors. Taking into consideration the fact that plants were possessed of numerous and varied therapeutic properties, he conceived the idea that the aggregate medicinal value of plants resided not in *one*, but in *several and distinct* proximate principles. Upon examining the ordinary preparations termed "concentrated," together with the methods employed for obtaining them, he soon ascertained that they were fractional and imperfect, consisting of isolated resin, resinoid, and alkaloid principles, as the case might be, and representing only in part the therapeutic constituents of the plants from which they were severally derived. As many of the preparations represented to be "*the* active principle" of certain plants were *insoluble* "resins" and "resinoids," and whereas the plants were known to yield *soluble* medicinal principles to water, additional evidence was afforded that some one or more of their active constituents were overlooked and lost. Furnished with this evidence, his investigations took a new direction, and their results are now laid before the profession. The existence of a multiplicity of active medicinal constituents in the same plant was correctly demonstrated, and two new classes of proximate principles, the *neutrals* and *muci-resins*, discovered and added to the list of those already known. These principles we have the honor of being the first to describe and introduce to the profession. To him belongs the credit of being the first to advance the idea of combining *all* the proximate medicinal constituents of a plant in one preparation—the first to make and announce to the profession correct chemical analysis of chemical plants, and the first to caution them against the unreliable character of extracts, syrups, and other of the ordinary preparations of the day.

While laboring faithfully during the past six years to advance the interests of organic chemical science, he has been none the less diligent in the discharge of the arduous duties of his profession, testing in clinical practice those preparations which his scientific skill had succeeded in bringing to a state as near perfection as possible, thus becoming a guarantee to the profession of the character of the remedial agents furnished. All preparations offered to the profession, emanating from his establishment, has been thoroughly tested in practice, unless explicitly stated to the contrary. Numerous improvements have been made from time to time, and "progress" is the rule of action with this gentleman.

All the preparations manufactured at his establishment are unlike those of any other manufacturer. The powdered preparations, as well as the concentrated tinctures, command the confidence and approbation of the profession. They are definite, reliable, and uniform in medical strength, portable, not liable to change, and convenient of administration. The concentrated tinctures are a peculiar feature in the improvements made by this gentleman in pharmaceutical science, of which we have already given a history. Every drop is of uniform therapeutic strength, and invariably represents a positive and definite amount of active principles.

To this gentleman and his co-laborers in the field of organic chemistry, belongs the credit of being the first to discover, describe, and introduce to the profession all but **two of the concentrated preparations** enumerated in this work.

# SENECIN.

Derived from *Senecio Gracilis*, Nat. Ord.—*Asteraceæ.*
Sex. Syst.—*Syngenesia Superflua.*
Common Names—*Life Root, Cough Weed, Waw Weed, Unkum, Female Regulator,* etc.
Part Used—*The Plant.*
No. of Principles, *two,* viz.: *resinoid* and *neutral.*
Properties—*Diuretic, diaphoretic, emmenagogue, febrifuge, expectorant, pectoral, alterative* and *tonic.*
Employment—*Amenorrhea, dysmenorrhea, menorrhagia, hysteria, gravel, strangury, chlorosis, dropsy, dysentery, gonorrhea, coughs, colds, loss of appetite, debility,* etc.

SENECIN is an elegant and efficient remedy, and one which admits of a wide range of application. It is deservedly held in high repute in the treatment of the various affections peculiar to females. From the fact of the plant having been successfully employed in domestic practice for *regulating* menstrual derangements, it derived one of its common names, that of *Female Regulator.*

Senecin, either alone or combined with other positive medical agents, has proved eminently successful in the treatment of amenorrhea. It is usually exhibited in doses of from two to five grains, three times per day. When the obstruction has arisen from cold, this remedy, in connection with warm alkaline pediluvia, is generally sufficient. If it be desirable to increase its diaphoretic effect, it may be advantageously combined with Asclepin. We employ the following formula:

℞.
 Senecin,
 Asclepin aa. grs. ij.

To be given at a dose, and repeated twice or thrice a day. When the affection is uncomplicated, we know of no remedy more generally reliable than the above. It operates kindly, and without excitement, and the catamenial flow is restored in a manner so natural that the patient is scarcely aware of being under the influence of medicine. Should the case prove obstinate, we administer a cathartic dose of Podophyllin at or near the usual time for the appearance of the menses, or whenever the system manifests a desire to restore this secretion. We seldom or never employ the Podophyllin alone, hence we resort to such combinations as the existing condition of the system may indicate. For the present purpose we usually give

 Podophyllin,
 Asclepin aa. grs. ij.

This may be generally given at bed time; but it is sometimes better to administer it as soon as any of the usual symptoms preceding the return of the menses are felt.

When the affection occurs in patients of a peculiarly nervous constitution, we combine the Senecin with Caulophyllin, as follows:

℞.
 Senecin,
 Caulophyllin, aa. ℈

Mix and divide into ten powders. Of these one may be given twice or thrice a day, at the option of the practitioner. By this combination we increase the emmenagogue property of the Senecin, and at the same time gain the anti-spasmodic effect of the Caulophyllin, which exerts a most desirable influence when this affection is accompanied with a convulsive tendency. The Viburin may be substituted for the Caulophyllin, and in some cases will answer a better purpose. If we require a more energetic relaxant and anti-spasmodic, we employ the Gelsemin. It may be substituted for either of the above, or may be combined to meet special indications, as in the following formulas. As an adjunctive, we have always found it valuable:

℞.
    Senecin grs. XXIV
    Gelsemin grs. IV

Mix and divide into eight powders. One of these may be given once in four hours. The quantity of Gelsemin may be increased or diminished according to the susceptibility of the patient's system to its influence. The repetition of the doses must be governed by the same considerations. As a more efficient combination still, the following may be employed:

℞.
    Senecin,
    Caulophyllin aa. grs. XX,
    Gelsemin grs. V.

Mix and divide into ten powders. Administer same as above. These combinations will be found very useful in controlling all spasmodic manifestations accompanying simple uncomplicated amenorrhea. But the Senecin should be employed alone in all cases where the above combinations are not positively indicated.

When complications exist or the case has become chronic, auxiliary remedies will be needed. These will depend, in each case, upon the existing necessity. If the liver be deranged in its functions, the prompt administration of Podophyllin or some other chologogue should precede all other treatment. If the biliary obstruction be slight, Leptandrin, Juglandin, or Irisin may be sufficient. If constipation be an attendant symptom, measures must be employed to obviate it. For this purpose from one fourth to one grain of Podophyllin, triturated with Asclepin, in the proportion of one to four, may be exhibited every night, or every second night. As a general thing we prefer to administer Podophyllin at night, and independent of whatever general remedies we may be employing, finding that it operates more kindly and pleasantly when thus exhibited.

We have derived equally happy effects from the employment of Senecin in the treatment of dysmenorrhea. The most beneficial results are obtained by exhibiting it during the intermenstrual period. It acts as a special tonic upon the uterine system, invigorating the menstrual function, and restoring equilibrium of action. For this affection it may be given in doses of from two to five grains two or three times a day, and alternated with Helonin. Or the two may be combined, as follows:

℞.
    Senecin grs. XX.
    Helonin grs. X.

Mix and divide into ten powders. This we have found to be a valuable combination. If the menstrual secretion be profuse, Trilliin should be substituted for the Helonin. If the secretion be scanty, Macrotin or Baptisin may be employed. Below we give our usual formulas:

℞.
    Senecin,
    Trilliin aa. grs. XVI.

Mix and divide into eight powders. These are to be used during the intermenstrual period, when the flow is immoderate.

℞.
    Senecin grs. XXIV.
    Macrotin grs. IV.

Mix and divide into eight powders. Or

℞.
    Senecin grs. XX.
    Baptisin grs. X.

Mix and divide into ten powders. Either of the above formulas will answer a good purpose when the secretion is defective. The above prescriptions are designed to constitute the radical treatment, while special symptoms must be met with such auxiliary measures as the circumstances of the case may demand.

We have been equally successful with the Senecin in the treatment of menorrhagia. It may seem somewhat paradoxical to the reader that we should prescribe the same remedy in what are generally conceived to be opposite conditions of the system. Thus amenorrhea and menorrhagia as supposed to indicate the necessity of remedies possessing dissimilar therapeutic properties. Let us look for a moment at the condition of the two cases. In each instance there is admitted to be deranged action. This disturbance of the physiological condition in either case is simply a loss of equilibrium. In the one case the functions are suppressed, and there is no secretion. In the other case there is a relaxed or enfeebled condition, and the secretion is profuse. We say secretion, but that is not the proper term. The act of secretion is purely a physiological phenomenon, accompanying, preserving, or restoring a normal condition. Profuse and active discharges are hardly to be looked upon in the light of a secretion, but rather as a sort of leakage, an indiscriminate outpouring of the constituents of animal fluidity. Secretion is the act of separating. As applied to the animal economy, it means to imply the process whereby a separation is effected between the vital and the morbid materials of the organism, the retention of the former, and the expulsion of the latter. It is not to be supposed that the system would reject any materials not yet become effete or useless, as such a proceeding would argue a prodigality and disposition to waste not at all in harmony with the wisdom displayed in its organization. Yet we find that these profuse secretions, so called, are a mixture of both the healthy and vitiated constituents of the body, and that the escape or flow is followed by exhaustion, impoverishment, and debility. This would certainly not be the case were the morbid materials only separated and expelled. Perspiration induced by exercise or vegetable diaphoretics is neither exhausting, or debilitating; but nightsweats, so called, are depleting and impoverishing in their effect. The latter are not the result of increased secretion, but are transudations resulting from a relaxed and enfeebled condition of the capillary vessels of the surface. The power to secrete is wanting, hence both the good and bad materials of the blood are allowed to run to waste through the unguarded portals of the skin. So in amenorrhea and menorrhagia. In both cases the power to secrete is wanting. In the one case it is suppressed in consequence of the interposition of certain obstructions. In the other case we have an illustration of that condition which has been designated by the term of *vis inertia*, or a complete passivity of the vital forces. Now it is evident that in either condition it is necessary to restore the secreting power, simply to recall and re-establish the functional equilibrium of the organs. No matter in which direction the scale may be turned, if we can but restore and equalize the functional activity of the parts, we shall effect a cure. For this purpose we employ the Senecin, simply because it possesses the power of recalling or restoring lost or healthful action. This then explains the seeming paradox of giving the same remedy in dissimilar derangements of the same organ. We shall have occasion to refer again to this subject in treating of other of the concentrated medicines. The plan of seeking to devise a different remedy for every variation in the manifestations of diseased action we deem to be erroneous, and calculated to confuse and render too complex the art of prescribing.

Our usual method of employing Senecin in the treatment of menorrhagia is the same as in the preceding cases. We rely upon it as a radical measure, while special symptoms are met as they arise.

Chlorosis is another of those incidental female affections in which the Senecin will be found an excellent remedy. In view of its alterative and tonic properties, it is peculiarly serviceable when chlorosis occurs in a strumous diathesis. In these cases it may be advantageously combined with other alteratives, as the Ampelopsin, Alnuin, Stillingin, Chimaphilin, &c., or with more decided tonics, as Cornin, Hydrastin, Menispermin, etc.

In anemic habits, the Senecin may be advantageously combined with the different preparations of Iron. Thus in some forms of chlorosis and amenorrhea, we may prescribe the following:

℞.
    Senecin ʒ ss.
    Iron by Hydrogen grs. VI.

Mix and divide into twelve powders. Dose, one, morning and evening. The quantity of Iron may be increased if deemed necessary. If constipation be an accompanying symptom, we may vary the prescription thus:

℞.
    Senecin,
    Leptandrin aa. grs. XX.
    Iron by Hydrogen grs. V.

Form a mass with mucilage of gum arabic and divide into ten pills. Dose, one, twice or thrice a day. The above will be found excellent for prolapsus uteri, when of an asthenic character. When the disturbance of the nervous system is considerable, and the symptoms verge on hysteria, we employ the Valerianate of Iron. It will answer the double purpose of relieving the anemic habit and allaying nervous excitability.

℞.
    Senecin ʒ ss.
    Valerianate of Iron grs. X.

Mix and divide into ten powders. Exhibit one morning and evening. The same will be found useful in chorea. When suppression occurs in females advanced in life, and when there are symptoms of a preternatural wasting of the tissues, we substitute the phosphate of Iron.

℞.
    Senecin grs. XXIV.
    Phosphate of Iron grs. VIII.

Mix and divide into eight powders. Dose same as above.

Senecin is valuable in the treatment of dropsy, not so much on account of its diuretic power as on account of its alterative and tonic properties, by reason of its exciting the glandular system to healthful action. The same may be said in relation to its employment in the treatment of gravelly affections.

In gonorrhea it manifests a decided sanative power. It may be employed alone, alternating with such other remedies as the features of the case may indicate, or it may be combined with other alteratives.

℞.
  Senecin,
  Stillingin ........................... aa. ℈ ij.
Mix. Dose, two to five grains three times per day.

℞.
  Senecin .............................. ʒ ss.
  Phytolacin ......... ................ grs. XV.
Mix. Dose, from two to four grains three times per day.

℞.
  Senecin .............................. ʒ j.
  Irisin.................................. ʒ ss.
Mix. Dose same as above.

℞.
  Senecin .............................. ʒ j.
  Corydalin .......... .................. ℈ j.
Mix. Dose, two to five grains. These formulas will be found equally serviceable in the treatment of syphilis. Other combinations may be effected when indicated. Thus if scalding of the urine be a troublesome symptom, Populin will be appropriate. If chordee be present, Stillingin is contra-indicated. Lupulin is then proper. Other agents may be added to the formulas given at the option of the practitioner, but we have found the simple combinations best, and prefer to use the auxiliary remedies separately.

Senecin has gained some repute in the treatment of dysentery, but our own experience of its value in that disease is limited. Our observation of its action in other diseases inclines us to the opinion that it would be mainly useful in the convalescing stages as a tonic.

In coughs, colds, and other complaints of the chest, Senecin is one of the most valuable remedies we possess. It is especially serviceable in mucous coughs. Either alone, or combined with Asclepin, Prunin, Hyosciamin, Lycopin, &c., it will seldom disappoint expectation.

℞.
 Senecin.
 Asclepin. .......................aa. ℨss.

Mix. Dose, two to four grains once in four hours. Serviceable when expectoration is difficult, skin dry, and system feverish.

℞.
 Senecin,
 Prunin aa ........................... ℨss.

Mix. Dose same as above. Useful when expectoration is free and tonics are indicated.

℞
 Senecin............................. ℨss.
 Hyosciamin ........................grs. ij.

Mix thoroughly, and divide into sixteen powders. Of these, one may be given once in from two to four hours. Excellent when pain is experienced in any part of the chest. Also, when the cough is troublesome at night. If there be symptoms of hemoptysis, the following will be the best combination.

℞.
 Senecin,
 Lycopin......................... aa. ℈j.

Mix and divide into ten powders. One may be given every three hours.

Taking into consideration the therapeutic properties already possessed by the Senecin, the practitioner may readily effect combinations with other agents calculated to simply augment each or either of the properties, or to increase their number, or to suppress the action of one or the other. Its range of application may thus be extended, although the remedy should be employed for its own peculiar merits.

## CONCENTRATED TINCTURE SENECIO GRACILIS.

This preparation of the Senecio is by some preferred to the Senecin. We are in the habit of employing it in cough mixtures, and in various ways. It is convenient for the practitioner when he wishes to leave medicine with the patient, or when sending medicine to a patient at a distance, as the labor and necessity of dividing it into separate doses is thereby obviated. Two drops of the Con. Tinc. represent one grain of Senecin, therefore the proper dose is easily estimated.

In remedial value it is fully equivalent to the Senecin, and may be employed in all cases where that remedy is indicated. In the treatment of amenorrhea, the following will be found valuable. The dose, and frequency of the repetition, must of course be regulated by the requirements of the case; we can only approximate it.

℞.
    Con. Tinc. Senecio Gracil.
    Con. Tinc. Gelseminum................aa. ℨij
Mix. Dose from five to ten drops three times per day.

For strangury and gravelly affections, we employ the following:

℞.
    Con. Tinc. Senecio Gracil.
    Con. Tinc. Eupatorium Purpu........... aa. ℨss
Mix. Dose four to eight drops, repeated once in from two to four hours, according to circumstances.

For hysteria, painful menstruations, etc., the following will be found excellent:

℞.
 Con. Tinc. Senecio Gracil.
 Con. Tinc. Scutellaria Later............ aa. ℨij.
 Con. Tinc. Hyoscyamus.................... ℨj.

Mix. Dose from five to fifteen drops, repeated once in from two to six hours, according to the urgency of the symptoms.

This will be found a reliable remedy for relieving pain and procuring rest in the above affections.

In the secondary stages of dysentery, after the secretive action of the liver has been corrected, and the inflammatory symptoms have measureably subsided, the following prescription will be found excellent for restraining and giving tone to the bowels:

℞.
 Con. Tinc. Senecio Gracil ................ ℨij
 Con. Tinc. Rhus Glab .................... ℨj.

Mix, and give from four to eight drops every two to four hours. If much prostration or sinking of the vital powers be present, the value and efficiency of the prescription will be materially enhanced by the addition of ℨj Con. Tinc. Xanthoxylum Frax. With this addition, it will prove an excellent remedy for cholera infantum, the morbid secretions having been first removed.

# ASCLEPIN.

Derived from *Asclepias Tuberosa*, Nat. Ord—*Asclepiadaceæ.*
Sex. Syst.—*Pentandria Digynia.*

Common Names.—*Pleurisy Root, White Root, Wind Root, Colic Root, Butterfly Weed,* etc.

Part Used—*The Root.*

No. of Principles, *two,* viz.: *resinoid* and *neutral.*

Properties—*Alterative, anti-spasmodic, carminative, diaphoretic, diuretic, expectorant, laxative* and *tonic.*

Employment—*Fevers of every type, pneumonia, croup, peritonitis, pleuritis, rheumatism, colic, colds, coughs, hepatic derangements, constipation, hooping cough, hysteria, amenorrhea, dysmenorrhea, leucorrhea, menorrhagia, and in inflammatory diseases of whatever type.*

No other remedy with which we are acquainted is so universally admissible in the treatment of disease, either alone or in combination, as the Asclepin. In fact, we can think of no pathological condition that would be aggravated by its employment. It expels wind, relieves pain, relaxes spasm, induces and promotes perspiration, equalises the circulation, harmonises the action of the nervous system, and accomplishes its work without excitement, neither increasing the force or frequency of the pulse, nor raising the temperature of the body. It is of especial service in the treatment of affections involving the serous membranes, as pleuritis, peritonitis, etc. The remarkable efficacy of the plant in the cure of pleurisy, for which purpose it has been employed for many years in domestic practice, has earned for it the common name of *Pleurisy Root*. In like manner it earned the appellation of *Wind Root* and *Colic Root*, having been found reliable as a carminative and anti-spasmodic.

In order that the full value of the Asclepin may be realized in the treatment of all febrile complaints, it must be exhibited in *full* doses and repeated sufficiently often to induce and maintain free diaphoresis. The usual dose of the Asclepin is from ONE to FIVE grains, but when there is high febrile excitement we commence with TEN grain doses, repeating every one or two hours until the system is brought under its full influence, and then diminish to from TWO to FIVE grains every two hours, or sufficiently often to secure the desired effect, that is, to sustain the diaphoretic action. It may accompany any other remedies without interfering with their specific properties, enhancing rather than retarding the action of such auxiliaries as may be used in connection with it.

Flatulent colic is quickly relieved by administering from FIVE to TEN grains of Asclepin every twenty minutes until the spasm is relaxed and the wind expelled. Relief will be more prompt if the remedy is administered in warm water. Cramp in the stomach will generally yield to the same prescription. The usual manner of exhibiting Asclepin in pleuritis is the same as in all febrile affections. Free perspiration must be induced and maintained for from twelve to twenty-four hours, or sufficiently long to overcome the local congestion. We have found it to act remarkably well in combination with Cypripedin.

℞.
 Asclepin............................... ʒj.
 Cypripedin ........................... ℈j.
 Aqua fervens ........................ ℥ IV.

Dose two teaspoonsfull every thirty minutes until perspiration is induced, then once in one or two hours as may be necessary to maintain the action. We have seen some very severe attacks of the pleurisy cured by this prescription alone. No depletion accompanies this treatment, and the patient is at once restored to his usual health. When a more active combination is needed, as for instance when there is excessive arterial excitement, we give the following:

℞.
  Asclepin............................. ℨ ss.
  Aqua fervens...................... ℨ IV.
  Con. Tinc. Veratrum Viride.............. gtt. XXX.

Dissolve the Asclepin in the water and add the Tinc. of Veratrum. Give two teaspoonsful every hour until the patient is brought under the full influence of the remedy, then repeat at intervals of two hours, or sufficiently often to keep the arterial excitement under control. If nausea arise, omit until it has subsided, then resume as before. Of course, we cannot name the precise dose, nor regulate the frequency of repetition for every case. The patient may require more or less than the dose we have advised, but all that is necessary is to give sufficient to produce the specific effect of the remedy, and to maintain the action until the disease is overcome. If nausea and even vomiting take place, no disadvantage will accrue, but, on the contrary, when the stomach is loaded with phlegm or other matters, will generally prove decidedly beneficial.

In the treatment of exanthematous fevers, of whatever type, we invariably employ the Asclepin. No remedy with which we are acquainted exercises so salutary an effect in these cases as the Asclepin. Its employment is admissible at any and all stages. It excites a kindly depurative action on the part of the cutaneous exhalents, and favors the development of the eruption. In the treatment of scarlatina it is of eminent service. Mild cases of scarlatina, rubeola, varicella, etc., are manageable with this remedy alone, and seldom is any auxiliary treatment necessary. When more active treatment is demanded, the formula given above will be found reliable. We seldom find any other medicines necessary in the treatment of scarlatina in this climate, except the occasional administration of a dose of Podophyllin.

Asclepin is an invaluable adjunctive in the treatment of many chronic diseases. From the fact of its exercising a peculiar influence upon the serous membranes, it proves a valuable remedy for chronic pleuritis, in which complaint it is most advantageously combined with Sanguinarin.

℞.
    Asclepin............................. ℈j.
    Sanguinarin.......................... grs. IV.

Triturate well together, and divide into ten powders. Exhibit one three times per day. This treatment, in connection with alterative doses of Podophyllin, will prove successful in a majority of cases. In obstinate cases, counter-irritation may be resorted to in connection with the above remedies. For this purpose the following will be found excellent:

℞.
    Ol. Stillingia Sylvat................... ʒ ij.
    Spts. Vini............................ ℥ IV.

Bathe the affected parts night and morning. Or the following:

℞.
    Ol. Stillingia......................... ʒ ij.
    Ol. Lobelia........................... ʒ ss.
    Spts. Vini............................ ℥ IV.

Apply same as above. This is excellent when it is desirable to produce relaxation. If a more stimulating application is indicated, we vary the formula, thus:

℞
    Ol. Stillingia......................... ʒj.
    Ol. Capsicum......................... gtt. X vel XX
    Alcohol .............................. ℥ ij.

This is a powerful stimulant and counter-irritant, and will be found eminently serviceable in arousing a proper action of the skin.

In all diseases accompanied with a dry skin, unequal circulation, feeble respiration, a tardy action of the renal functions, flatulence, constipation, or viscidity of the secretions, Asclepin will prove a most reliable remedy, either alone or in combination with other agents. For the removal of hepatic obstructions, it may be advantageously combined with either of the following agents: Podophyllin, Leptandrin, Juglandin, Euphorbin, Irisin, Phytolacin, or Apocynin. In the treatment of Rheumatism, with Macrotin, Sanguinarin, Xanthoxylin, Phytolacin, Stillingin, or Rumin. For the cure of Chronic coughs, with Prunin, Cerasein, Senecin, Lupulin, or Sanguinarin. For hemoptysis, with Lycopin, Trilliin, or Eupatorin,. (Purpu.) It is true that it may be considered as simply an auxiliary to some of the above mentioned remedies, yet we know full well that their efficacy is materially enhanced by the modifying action of the Asclepin. The only difficulty is, that Asclepin is too frequently looked upon as a simple and inadequate remedy, which needs must be combined with some more potent agent, and hence it is too seldom employed alone. Were more confidence reposed in its therapeutic worth, it would be found that no one agent manifesting so little excitement in its operation is capable of successfully meeting so great a number of indications. Possessing alterative, laxative, and tonic properties, it is exceedingly valuable in the treatment of some forms of indigestion, increasing the appetite, promoting digestion, and removing constipation. In the cure of hooping cough, it is with us a favorite remedy. From THREE to FIVE grains may be given four times a day. We usually form a solution with warm water. If the cough is violent or spasmodic, we add from FIVE to TEN drops of the Wine Tinc. of Lobelia to each dose. We sometimes use the Asclepin in connection with Hydrocyanic Acid.

℞.
  Hydrocyanic Acid....................gtt. X.
  Water............................. ℥ IV.

Dose one teaspoonful three times a day. At the same time we give the Asclepin in sufficient quantities to maintain a gentle diaphoresis. No other plan of treatment that we have ever seen devised has proved so uniformly successful as the above, cutting the disease short with remarkable certainty.

Asclepin is one of the most valuable remedies in the advanced stage of phthisis pulmonalis that we have ever employed. It overcomes the viscidity of the secretions, promotes expectoration, abates febrile excitement, and by promoting the cutaneous exhalations, lessens the cough. And all this it does so kindly that the patient is surprised and delighted at the degree of comfort ensured by so mild and pleasant a remedy. Its action is so different from the Diaphoretics usually employed, that its employment is always admissable, and will not interfere with the action of such anodynes or sedatives as the physician may have occasion to administer.

In the treatment of hysteria, amenorrhea, and other diseases incident to females, the Asclepin proves a remedy of much utility. We shall frequently refer to it when treating of other remedies, as no other agent will admit of so frequent and promiscuous combination. It may be thought that we are too sanguine in our advocacy of the virtues of the Asclepin, but we rely upon a verdict in favor of the truthfulness of our estimate from all who have had a similar experience with ourselves.

In the management of dysentery, the Asclepin will be found an indispensable auxiliary when once its real value is understood. Diaphoretics are always indicated in that disease, and none will be found more valuable than the Asclepin. We have frequently known a single dose to cure a severe diarrhea. When arising from cold, the cure is almost certain. In the treatment of cholera infantum we have found the Asclepin a highly useful remedy.

As stated in the first part of this work, we employ the Asclepin as a substitute for sugar, etc., in triturating the more active concentrated medicines. Among these we may enumerate the Veratrin, Hyoscyamin, Digitalin, Sanguinarin, Podophyllin, and Gelsemin. We know of no indication in which the Asclepin would be inadmissable; neither will it interfere in suppressing the therapeutic action of either of these remedies.

On the contrary, it will increase their activity, and, by rendering them more diffusible, insure a more kindly operation. We know of no combining agent so generally appropriate, or which exercises a more desireable modifying influence over the Podophyllin than the Asclepin. **The combinations will be noticed in connection with each agent.**

# GELSEMIN.

Derived from *Gelseminum Sempervirens.*
Nat. Ord.—*Apocynaceæ.*
Sex. Syst.—*Pentandria Digynia.*
Common Names.—*Yellow Jessamine, Wild Jessamine, Woodbine,* etc.
Part Used—*Bark of the Root.*
No. of Principles, *three,* viz., *resinoid, neutral* and *alkaloid.*
Properties—*Febrifuge, nervine, anti-spasmodic, relaxant, alterative, emmenagogue, parturifacient,* and *narcotic.*
Employment—*Fevers, pneumonia, pleuritis, rheumatism, hysteria, dysmenorrhea, amenorrhea, gonorrhea, chorea, spermatorrhea, epilepsy, paralysis, after pains, convulsions, and to expel worms.*

ALTHOUGH comparatively a new remedy, the Gelseminum has rapidly gained the approbation and confidence of the profession. We are firmly of the opinion that not one half the true value of the Gelseminum is understood, yet sufficient is already known to render it a most welcome addition to the Materia Medica. As the plant is possessed of most positive and active therapeutic powers, it is important that its pharmaceutical preparations should ensure a definite and uniform standard of medicinal strength. Such a *desideratum* has been secured in the preparation now under consideration. The three active principles of the plant have been isolated and recombined, and form a beautiful and convenient powder. Numerous attempts have been made to isolate the active principles of the Gelseminum, so as to secure them in a powdered form, but this result has only been accomplished at the laboratory of B. Keith & Co. The thanks of the profession are due, in this instance, as in many others, to the indomitable energy and skill of this firm, in having so faithfully rendered us a concentrated equivalent of the plant.

Gelsemin is deservedly entitled to the appellation of *positive* medical agent, being possessed of specific and positive therapeutic properties, uniform in strength, and capable of preserving its properties unimpaired for an unlimited period of time.

The average dose of the Gelsemin is HALF a grain. But owing to constitutional peculiarities, the dose will vary from ONE-FOURTH to TWO grains. Fevers and inflammatory diseases generally afford a spacious field for its employment. Its peculiar influence over the nervous and circulating systems justly entitles it to be called both nervine and febrifuge. A knowlege of the peculiar febrifuge power of this remedy, has ushered in a new era in the treatment of febrile diseases. Fevers of almost every type may be controlled in from six to eighteen hours. In order to reap the full utility of the remedy, it must be given in sufficient doses to produce its constitutional effects, and the patient kept fully under its influence until the symptoms are completely subdued. The effects referred to are dimness of vision, double-sightedness, inability to open the eyes, and, when carried beyond this, complete prostration of the muscular system. But it is seldom necessary to carry the administration of the remedy to the production of the latter influence. It is sufficient in a large majority of cases to produce a slight dimness of vision, and to continue the remedy with such doses and frequency of repetition as will maintain a uniform degree of action at this point. In many cases it will be expedient to reduce the dose to just below the production of this effect. Even when the remedy has been carried to the production of complete bodily prostration, we have never known any permanently injurious effects to remain. These symptoms will all pass off in a few hours, leaving the patient refreshed and positively invigorated, rather than leaving, as might be expected, any symptoms of exhaustion or debility. It is always best to explain to the patient and attendants the nature of the symptoms likely to arise when this remedy is exhibited, otherwise unnecessary alarm may be excited, and, as is frequently the case, the nurse, in the absence of the physician, will administer stimulants, and so defeat the action of the remedy. In the treatment of pneumonia, it is sometimes necessary to keep the patient under the full influence of the Gelsemin, that is, to the production of dimness of vision or double-sightedness, for four or five days. If this be not done, the disease will progress unconquered, and the patient be lost. Some division of opinion exists as to whether the Gelsemin has a narcotic property. We should think that a very slight experience would be sufficient to decide this question. When the patient is brought fully under its constitutional influence, the symptoms are so marked that we cannot conceive how the remedy should be deemed otherwise. On attempting to move about, the patient appears as if intoxicated, the muscles refuse to obey the mandates of the will, while the head is dizzy, and the senses confused. In some respects the symptoms much resemble those produced by Strammonium, and in like manner pass off as soon as the remedy is discontinued. At other times the patient appears as if under the influence of alcohol, and evinces a decided disinclination to motion, and a tendency to sleep, from which he awakes feeling invigorated and refreshed.

In some instances, in the treatment of fevers, it is best to precede the employment of the Gelsemin with a cathartic dose of Podophyllin. In general, if we find that Podophyllin is indicated, we administer it in combination with Gelsemin:

℞.
    Podophyllin,
    Asclepin .......................aa. grs. ij.
    Gelsemin ........................gr. j.

Or,

℞.
    Podophyllin,
    Leptandrin......................aa. grs. ij.
    Gelsemin........................grs. j.

With a single dose of either of the above formulas we have frequently arrested typhoid and other fevers in the forming stages, so completely as to render further medication unnecessary. A more powerful combination is the following:

℞.
    Podophyllin,
    Euphorbin......................aa. grs. ij.
    Gelsemin........................gr. j.

This will prove an emeto-cathartic dose, and we have frequently arrested severe attacks of fever, rheumatism, and pneumonia, by exhibiting it in the forming stages. This may be deemed heroic treatment, but in the section in which we write, it answers our purpose, and that is just what we desire of every remedy. If any fever remain after the operation of the above, we follow with the Gelsemin until it is controlled. Asclepin may always be advantageously exhibited in connection with the Gelsemin. This is particularly the case in pneumonia, scarlatina, and eruptive fevers generally.

Acute rheumatism will frequently yield to the Gelsemin, particularly if the system has been properly regulated by the previous exhibition of Podophyllin. But it must be remembered that the Gelsemin is not a specific, and that many constitutions will not bear it at all, while others seem to be completely fortified against its impressions altogether, experiencing no influence from it whatever. In such cases we must rely upon the Veratrin.

It is in the treatment of female disorders that we find the Gelsemin peculiarly serviceable. Amenorrhea will frequently yield to Gelsemin when administered in HALF grain doses three times a day. Hysteric convulsions are also readily controlled with it. For relieving the pains of dysmenorrhea, we know of no single remedy equal to it. We give from ONE HALF to ONE grain every two hours. If it fails alone, we give the following:

℞.
    Caulophyllin,
    Viburin..........................aa. grs. XX
    Gelsemin..........................grs. V

Mix, and divide into ten powders. Give one every two hours. If the pain is very severe, repeat every hour. This is without exception the most efficient remedy for the relief of pains accompanying menstruation with which we are acquainted. When caused by functional derangement, we deem it a specific. We have earned the gratitude of many sufferers by the employment of the above. It is equally efficacious in relieving the pains occurring after parturition. Neuralgia will also often yield to the same prescription. In connection with suitable tonics, Gelsemin will be found of great service in the treatment of chorea. The tonics employed should be of an anti-periodic character, such as Cornin, Cerasein, and Iron.

Gelsemin has gained considerable repute in the treatment of gonorrhea. We have employed it for some three years past in that disease, but have never relied upon it exclusively. Our principal object in employing it is to overcome the urethral inflammation, and prevent chordee, and for these purposes we have found it reliable. It may be given alone or in combination with alteratives. We usually administer it at bed time, finding that the patient is more apt to enjoy a quiet night's rest thereby. From ONE to TWO grains of the Gelsemin, or from TEN to TWENTY drops of the tincture may be given. While some patients are readily controlled by SIX or EIGHT drops, we have found some to require TWENTY-FIVE drops for the same purpose. We cannot say with certainty whether the Gelsemin possesses any specific alterative value in the above disease or not, but we believe it does, and in that belief we prescribe it in all the cases we are called upon to treat, as an auxiliary.

For spermatorrhea, in connection with tonics, we have found it of exceeding utility, In many cases it is better to administer the Gelsemin alone for a few days, or until a remission of the symptoms is induced, and then follow with tonics. Of the latter, Cerasein will be found most efficient. In some cases Lupulin, Hydrastin, or Cornin will answer a better purpose. At other times we combine the Gelsemin with tonics, as follows:

℞.
  Cerasein............................ ʒj.
  Gelsemin..........................grs. Vj.

Mix, and divide into twelve powders. Dose—one, three times per day. In some cases, double the above dose will be required. The formula given below we deem the most efficient that can be devised:

℞.
  Cerasein........................ ʒj.
  Lupulin..........................grs. XXIV.
  Gelsemin ......................grs. Vj.

Mix and divide into twelve powders, same as the above. As soon as the emissions are effectually checked, we omit the Gelsemin and continue the Cerasein and Lupulin for at least one month. When the affection arises from a badly cured gonorrhea, we direct injections of Chloride of Lime to the urethra.

℞.
  Chloride of Lime...................... ʒj.
  Water ..............................O.j.

Inject three or four times a day. If too strong, dilute. This treatment has cured some obstinate cases.

We have found the Gelsemin remarkably efficacious in some forms of convulsions. Not only will it control the spasms, but also effect, in many cases, a cure, as it is a direct tonic to the nervous system. The doses should be sufficiently large to bring the system under control, and as soon as a remission is fairly established, the dose should be diminished one-half, and continued as long as may be thought necessary. It is advisable, in some instances, to combine the Gelsemin with anti-periodics, as soon as a remission occurs, precisely as in the treatment of intermittent fever. Should the convulsions return, omit the tonic until another remission occurs. Tonics, however, will sometimes aggravate the disease, in which case the Gelsemin will answer a better purpose alone We have cured several cases of epileptic convulsions by occasionally exhibiting a dose of Podophyllin, with Gelsemin at night, and Cerasein during the day. We also direct that, if the patient be conscious of the approach of the fit, a dose of the Gelsemin be taken immediately, which will usually have the effect of preventing its recurrence. It is sometimes advisable to administer the Gelsemin two or three times a day, so as to keep the system continually under its influence. As soon as the disease is controlled, the doses of the Gelsemin may be diminished in frequency.

Hysteric convulsions, when not arising from displacement of the uterus, may also be controlled with the Gelsemin.

Some division of opinion exists in relation to the true action of this remedy upon the uterus. We have had considerable experience in the treatment of female disorders, and have used the preparations of Gelseminum quite extensively. For five years past we have employed it as a parturifacient, and with better satisfaction than any other remedy. We use it for the purpose of relieving cramps, or other spasmodic difficulties, vertigo, nervous irritability, wakefulness, and other symptoms accompanying gestation. We usually commence its employment about five weeks before the expected time of confinement, if not sooner indicated, and exhibit from ONE-FOURTH to ONE-HALF grain of the Gelsemin every other night, or from

FIVE to TEN drops of the Con. Tincture. The Gelsemin, however, will not agree with all constitutions, and we have met with some two or three cases in which we could not employ it. Where no such idiosyncracy exists, it will compose both the mind and body of the patient, and carry her safely and fully up to the completion of the period of gestation. It seems to prepare the system for the parturient effort, and labor is completed in an unusually short period of time. As soon as delivery is effected, and the secundines expelled, we give the patient from ONE-FOURTH to ONE-HALF grain of Gelsemin, or FIVE to TEN drops of the Concentrated Tincture. This quiets all nervous excitability, favors the contraction of the uterus, and acts as a prophylactic of febrile excitement. It must be borne in mind that Gelsemin is *narcotic*, and hence will not be admissable at all times. We have met with a few cases of pregnancy in which the Gelsemin was indicated, but owing to existing idiosyncracies it could not be employed. In some cases it will fail to produce the desired effect, without otherwise manifesting any impressions upon the system, simply failing to act all. In other cases it will produce considerable cerebral excitement, with a tendency to vertigo, and without relieving the symptoms for which it was administered.

Gelsemin is one of those medicines which are peculiarly governed in their action by the quantity administered. Thus in small doses it acts as a gentle stimulant and tonic to the nervous system, giving vigor and harmony of action; while in large doses it proves a powerful relaxant, completely prostrating the muscular system, and, by over stimulating the brain and nerves, produces irregular and disturbed nervous action. The opinion has been entertained by some, that the Gelseminum is capable of producing abortion, but our experience with it inclines us to the contrary belief. As before stated, when administered in small doses, it gently stimulates uterine contraction, but when given in large doses it will arrest the progress of labor with much certainty. Still we are unable to say that it will *not* produce abortion under some circumstances, although we have never seen any evidence of its power to do so, and we have administered it to females at all the different stages of utero-gestation.

Gelsemin has proved effectual in expelling intestinal entozoa, particularly the *ascaris lumbricoides* and *tricocephalus dispar*. The Gelsemin may be administered in ONE-HALF or ONE grain doses two or three times a day, as the patient will bear, for two or three days, and then followed with a brisk cathartic. Or it may be combined with Podophyllin, as in the following formula:

℞.
    Gelsemin............................grs. V
    Podophyllin..........................grs. X

Mix, and divide into ten powders. Exhibit one every night for three nights, then omit three nights, and repeat as before. If the bowels should not be sufficiently relaxed by the use of one of these powders daily, the quantity of Podophyllin may be increased, or an additional powder may be administered in the morning. Other formulas embracing the Gelsemin will be given under the head of Santonin.

Neuralgia, when arising from functional disturbances of the nervous system, is successfully treated with Gelsemin. From ONE-FOURTH to ONE grain of Gelsemin, or from FIVE to FIFTEEN drops of the Con. Tinc. may be given every two hours until relief is obtained, and then at longer intervals until the affection is broken up. We frequently form combinations of Gelsemin with other neuropathics, as Cypripedin, Scutellarin, Lupulin, Hyosciamin, etc., as may be indicated at the time. In many cases of neuralgia, the use of Gelsemin, or other remedies of its class, will prove of but temporary service unless accompanied with, or followed by a tonic of an antiperiodic character. The Gelsemin, however, possesses considerable anti-periodic power, and will prove more uniformly permanent in its action upon the nervous system than many other remedies of its class. Gelsemin may be combined with anti-periodics in the treatment of neuralgia, but we prefer to administer it alone until we have obtained a remission of the symptoms, and then follow with Cerasein, Cornin, Hydrastin, or

Quinine, either alone or combined with Iron, in such doses, and with such frequency of repetition as the circumstances of the case will justify.

It would be impossible for us to give a full and complete history of the range of employment of the remedy under consideration. Our experience in the use of this remedy has not been limited, yet we feel that we have but feebly portrayed its therapeutic value. It has proved reliable in our hands in fulfilling all the indications of disease we have mentioned, yet we do not, by any means, look upon it as a specific. In the absence of any idiosyncracy on the part of the patient forbidding its employment, it is a sure and effectual remedy in controlling febrile excitement. It was the first remedy introduced to the profession by which typhoid and other fevers could be completely controlled and subdued in from twelve to eighteen hours, thus disproving the statement that such types of disease must "run their course." That it is capable of doing this, we have but to refer to the corroborative experience of all who have understandingly employed it for this purpose. Giving tone and harmony of action to the nervous system, it proves an invaluable remedy in the treatment of all spasmodic affections.

The Gelsemin is a remedy not to be incautiously trifled with, and those adopting its use should commence with small doses until they learn by experience somewhat of its peculiar influences. Avoid combinations as much as possible, and rely rather upon alternation. In this way the true value of the remedy may be learned. The medium dose of the Gelsemin is HALF a grain.

# CON. TINC. GELSEMINUM SEMPERVIRENS.

This preparation of the Gelseminum is equivalent in therapeutic properties to the Gelsemin. It is prepared in accordance with the conditions of the method referred to in the first part of this volume, and possesses the advantage over all other prepared tinctures of this plant of being of UNIFORM medicinal strength.

The medium dose of this tincture is TEN drops. In many cases FIVE drops will produce the peculiar constitutional influences of the plant, while in other cases as many as THIRTY drops will be required. We are of opinion that the action of the tincture is in general more prompt than that of the Gelsemin, in consequence of its diffusible character. It is very convenient for combining with other tinctures, and for adding to solutions of other remedies. It also enables us to graduate the doses with much precision.

The tincture may be employed for all the purposes for which we have recommended the Gelsemin. In the treatment of febrile diseases, we employ it in connection with Asclepin, as follows:

℞.
  Asclepin ............................ ℨ ss.
  Warm water ........................ ℨ ij.
  Con. Tinc. Gelseminum ............ gtt. LX.

Dissolve the Asclepin in the water and add the Tinc. Gelseminum. Dose, from one to three teaspoonfuls once in two hours. This is a very convenient form of preparing it for administration in the above mentioned diseases, particularly when a continued use of the remedy is necessary, and when the physician cannot conveniently see the patient sufficiently often to superintend its exhibition.

We employ the tincture very frequently in the treatment of chronic diseases as a matter of convenience, as the patient is enabled to estimate the dose by the number of drops directed. In commencing the use of the tincture in chronic disease, we order what we consider to be rather less than a medium dose for the patient in hand, and direct that, if the peculiar constitutional impressions are not produced by that quantity, the dose be increased one drop at a time until the symptoms of dizziness or clouded vision are apparent, then to hold at that quantity, or reduce a drop or two, and thus continue.

Combinations are very readily effected with other of the concentrated tinctures when desired. Thus with Con. Tinc. Senecio as recommended under that head for amenorrhea. In the treatment of nervous affections it may be advantageously joined with Con. Tinc. Scutellaria.

℞.
    Con. Tinc. Gelseminum,
    Con. Tinc. Scutellaria ................aa. ʒj.

Dose, from FIVE to FIFTEEN drops.

For hooping cough, asthma, etc., joined with the Wine Tinc. of Lobelia, it will be found very beneficial.

℞.
    Con. Tinc. Gelseminum ................. ʒss.
    Wine. Tinc. Lobelia .................... ʒj.

Mix. Dose, FIVE to TEN drops once in three hours, or whenever the cough is troublesome.

Combined with the Con. Tinc. Apocynum, we have a very excellent remedy for the removal of ascaris vermicularis.

℞.
    Con. Tinc. Gelseminum.................. ʒj.
    Con. Tinc. Apocynum................... ʒss.

Mix. Dose, from SIX to TWELVE drops three times per day. After using the remedy for three days in this manner, if the bowels are not sufficiently relaxed, administer a dose of Podophyllin. This will generally prove most effectual in expelling those vermin.

For the removal of the ascaris lumbricoides, a useful com-

bination may be effected with the Con. Tinc. Chelone Glabra
℞.
    Con. Tinc. Gelseminum .................. ℨj.
    Con. Tinc. Chelone ..................... ℨij.
Mix. Dose, from FIVE to TEN drops three times per day, for three days, followed by a dose of Podophyllin, or some other cathartic. If the first trial should prove ineffectual, repeat in the same manner.

We have found the tincture beneficial as an outward application in various affections. Diluted with from four to eight parts of water, we have applied it with excellent results to erysipelatous inflammations. The parts should be kept covered with cloths wetted in the dilute tincture. It abates the local inflamation, and has a very soothing and pleasant influence. The same application has been found beneficial in inflammation of the eye, resulting from cold, as well as in purulent and other forms of opthalmia. Wash the eye with the dilute tincture, and then apply cloths wetted with it as above directed. Diluted in the same manner, and dropped into the ear, it will soften the accumulations of hardened cerumen, and relieve the ringing, roaring, and other disagreeable symptoms that result from deranged secretion.

We have found the Tinc. an excellent remedy for poisoning by the Rhus Rhadicans, and Rhus Toxicodendron, common names, poison ivy, and swamp or poison sumach. Dilute the tincture with from four to eight parts of water and apply as directed for erysipelas, keeping the parts constantly moistened with it. If there be any febrile excitement present, administer the tincture internally at the same time, in such doses, and with such frequency of repetition as the case will warrant. We have experienced the value of this remedy in our own person, and can recommend it as reliable. We also have the concurrent testimony of practitioners who have used it for the same purpose.

The dilute tincture is also beneficially applied to some forms of rheumatic swellings, neuralgic affections, etc. We frequently combine it with other bathing preparations.

The following is excellent:

℞.
  Soap Liniment .......................... ℨiij.
  Con. Tinc. Gelseminum ............... ℨj.

Mix. Bathe the parts freely, repeating every two or three hours, or apply cloths wetted with the mixture, covering with a dry bandage to prevent too rapid evaporation.

Many forms of skin diseases may be benefited and cured by the internal and external application of the tincture. For external application the above mixture will be found useful, or the tincture may be added to ointments, or mixed with other fluid applications.

# MACROTIN.

Derived from *Macrotys Racemosa*
Nat. Ord.—*Ranunculaceæ.*
Sex. Syst.—*Polyandria Di-Pentagynia.*
Common Names.—*Black Cohosh, Deer Weed, Rattle Root, Black Snake Root, Squaw Root,* etc.
Part Used—*The Root.*
No. of Principles, *three,* viz., *resinoid, alkaloid* and *neutral.*
Properties—*Alterative, anti-spasmodic, stimulant, diaphoretic, diuretic, expectorant, resolvent, nervine, emmenagogue, parturient, tonic* and *narcotic.*
Employment—*Amenorrhea, leucorrhea, dysmenorrhea, hysteria, chorea, chlorosis, to facilitate delivery, rheumatism, coughs, colds, asthma, hooping cough, phthisis, small-pox, croup, convulsions, epilepsy, neuralgia, scrofula, indigestion, prolapsus uteri, gonorrhea, gleet, spermatorrhea, intermittent fever, cutaneous diseases, bronchitis, laryngitis,* etc.

IT may be thought that we have awarded to the Macrotin a too liberal range of employment: but we can assure the reader that we write from positive data, and with the record of our own and cotemporary clinical experience before us. With this assurance we shall proceed to lay before the reader a history of its application in disease.

The alterative properties of this remedy are well marked, hence its utility in scrofula, cutaneous diseases, &c. We shall not assume to explain the manner of its operation in these cases, but confine ourselves to a history of results. We do not look upon it as a specific in disease, but as of great reliability in fulfilling specific indications. As with all other remedies possessing alterative properties, its successful employment is based upon certain conditions. Thus, in scrofula, we should correctly estimate the necessities of the system, and determine whether those conditions are present or not. As the remedy imparts a healthful stimulus to the digestive and nutritive functions, we should see that the elements of nutrition are supplied, in order that, if activity be given to the functions of nutrition, there be something upon which the action so aroused may expend itself. It is worse than useless to excite the nutritive apparatus of the system to action unless there be material to appropriate. Scrofula occurs mostly in patients whose systems are deficient in nitrogenous matters and iron, hence the latter are to be supplied as articles of diet or materials of sustenance and reparation, while the Macrotin will act as a motor-excitant, promoting the assimilation and appropriation of the sustaining and reparative material. By observing these conditions, the practitioner will find in the Macrotin a most excellent remedy for the treatment of the above named diseases. It exercises a remarkable influence over the nervous system, giving tone and harmony of action, and awakening its latent energies to healthful activity. This peculiar stimulant property is of great service in those cold and passive conditions which sometimes attend the development of strumous diseases. In such cases it proves a valuable adjunctive to other alteratives and tonics. It may be given alone and alternated with other appropriate remedies, or combined with such alteratives or tonics as are indicated. The medium dose of the Macrotin is HALF a grain. When given in small doses, it gently stimulates the nervous system, relaxes muscular spasm, allays pain, soothes the irritability of the system, reduces the force and frequency of the pulse and equalizes the circulation, and acts as a prophylactic of cerebral congestion. In over-doses it produces considerable cerebral disturbance, with vertigo, nausea, prostration, pain and fullness in the head, and an indefinable sense of aching in the joints. In its general influence, when taken in large quantities, it simlates the action of alcohol. An infusion of green tea or roasted coffee counteracts its impressions. We have never known any permanently injurious effects to follow the production of the above symptoms, yet in patients of a peculiarly susceptible organism we would advise caution in its employment.

In the treatment of amenorrhea, the Macrotin may be given in doses of from ONE-FOURTH to ONE grain, three times per day. In order to be effectual, it is generally necessary that the doses should be sufficiently large to produce the constitutional effects of the medicine in a slight degree. In many cases these symptoms will be limited to a slight sense of aching in the joints, and a peculiar electrical sensation extending throughout the entire system. At other times these peculiar sensations will be manifested only in the organs or parts diseased, as in the kidneys, liver, etc. In the treatment of the affection under consideration, the Macrotin may be alternated with such other medicines as the necessity of the case demands. Thus if it be desirable to increase its emmenagogue and tonic properties, it may be alternated with Senecin, Helonin, Baptisin, etc. The Macrotin may be exhibited for a few days, and then followed with either of the above remedies, or they may be alternated upon the same day. To avoid complexity, combinations may be formed. Thus to increase its tonic, stimulant, emmenagogue properties, as follows:

℞
    Macrotin  ·  ·  ·  ·  ·      grs. V.
    Senecin  ·  ·  ·  ·  ·  ·    Ɔj.

Mix, and divide into ten powders. Dose, one, three times per day. Or the following:

℞.
    Macrotin - - - - - - grs. VI.
    Helonin - - - - - grs. XVIII.

Mix, and divide into twelve powders. Dose, same as above. When laxatives are indicated, it is better to exhibit the Macrotin through the day, and the laxative at bed-time.

In the treatment of leucorrhea the Macrotin should be given in doses sufficiently large to produce the constitutional symptoms, and warm alkaline hip baths employed every day. In speaking of these complaints, we mean to be understood as referring to simple uncomplicated affections. When complications exist, the indications must be determined and met according to the individual characteristics of each case.

Dysmenorrhea is frequently relieved of its immediate painful character by administering from ONE-HALF to ONE grain of Macrotin every two hours, and permanently cured by continuing the remedy, in appropriate doses, during the intermenstrual period.

The spasms of hysteria, when not arising from actual displacement of the uterus, are easily controlled with the Macrotin. If there be prolapsus, inversion, or retroversion of the uterus, first replace it, then administer the Macrotin, and having quieted the immediate irritability, continue the remedy until the tone of the system is restored, and thus guard against such accidents in future.

The Macrotin possesses considerable anti-periodic power, hence will be found useful in the management of chorea. Exhibit in full doses, and alternate during the remissions with more decided tonics, such as Cornin, Cerasein, Hydrastin, Quinine, Iron, etc. If the Macrotin should not prove sufficiently anti-spasmodic, it may be joined with other remedies of its class. Among these may be enumerated Gelsemin, Viburnin, Cypripedin, Caulophyllin, and Veratrin.

In connection with Iron, Macrotin will be found valuable in the treatment of chlorosis. It must be borne in mind that Macrotin will increase the activity of those remedies with which it may be combined. This it does, not by actually increasing the medicinal power of the adjunctive, but by arousing the impressibility of the nervous system, and by promoting its absorption and diffusion. For the complaint above mentioned we may combine the Macrotin as follows:

℞.
  Macrotin . . . . . . grs. V.
  Iron by Hydrogen . . . grs. X.
Mix and divide into ten powders. Dose—one, twice a day. Under all circumstances the acidity of the stomach should be neutralised before exhibiting the Macrotin. Other preparations of Iron may be substituted for the above, as the Valerianate, Phosphate, Carbonate, etc.

 For promoting delivery, the Macrotin is deservedly held in high repute. It is indicated in all cases in which Ergot is usually employed, and we have the testimony of several eminent practitioners that it is not only equal, but preferable under all circumstances. When the uterine efforts are feeble and irregular, the Macrotin should be exhibited in doses of HALF a grain once in two hours. It is very important to not administer the remedy in too large doses, otherwise the object in view will be defeated. This is a general error in the employment of Ergot, overaction being quite too frequently produced. If the uterus be undilated, or undilatable, the use of the Macrotin should be preceded by the Wine Tinc. of Lobelia. We have been assured by those who have employed the Macrotin, that they would never again use Ergot, being satisfied that the former is quite as efficient, and, at the same time, much more kind and safe in its operation. It is the opinion of some that the Macrotin is inferior as a partus accelerator to the Caulophyllin; but both are good, and as neither are specifics, one may answer where the other fails.

 Macrotin is highly esteemed in the treatment of chronic rheumatism, in which complaint it is quite as reliable as any other single remedy. The patient must be brought under its full influence, and the remedy persevered with. In this complaint it is advantageously combined with Sanguinarin, Xanthoxylin, Stillingin, Irisin, Phytolacin, Rumin, etc.

℞.
  Macrotin ............................grs. V.
  Xanthoxylin......................... ℈j.

Mix, and divide into ten powders. Dose—one, three times per day. Diaphoretics are always of service in rheumatism, hence we employ the following combinations:

℞.
  Macrotin ............................grs. X.
  Sanguinarin .........................grs. V.
  Asclepin ............................grs. XL.

Triturate well together and divide into twenty powders. Dose —same as above. Or,

℞.
  Macrotin ............................grs. X.
  Phytolacin ..........................grs. XX.
  Asclepin ............................grs. XL.

Triturate and divide into twenty powders. Exhibit same as above. In this way we form combinations with other remedies suited to the case in hand. As a general thing the employment of these remedies in rheumatism should be preceded by the use of Podophyllin, and an occasional dose should be administered during the progress of the treatment.

Macrotin possesses well marked expectorant and diaphoretic properties, hence is valuable in the treatment of colds, coughs, incipient phthisis, etc. In these affections it may be either alternated or combined with Senecin, Asclepin, Prunin, Sanguinarin, or Lycopin.

In view of its anti-spasmodic and expectorant properties, the Macrotin has been found highly beneficial in asthma, hooping cough, and croup. As an expectorant, it may be employed with confidence whenever such a property is indicated. For asthma or hooping cough, it is excellent when joined with Eupatorin Purpu. or Apocynin, or Prunin, etc. In croup, after the urgent symptoms are all tyed, it is exceedingly beneficial as an expectorant. In all spasmodic affections of the respiratory system it is a reliable and valuable remedy.

The Macrotin has been highly recommended in the treatment

of small pox. Our experience of its employment in that disease has been somewhat limited, yet sufficient to give us a very high estimate of its value. We have exhibited it in a number of cases with obviously good effects. When administered during the febrile stage, it reduces the force and frequency of the pulse, allays cerebral excitement, equalises the circulation, and induces a gentle diaphoresis. We are satisfied that it will modify the violence of the symptoms, and deprive the disease of much of its malignancy. It is also of value in the treatment of other eruptive fevers.

Epilepsy has been much benefited by the use of Macrotin. It will usually induce a remission of the symptoms, although it may not prove sufficiently anti-periodic to prevent their recurrence. In such an event it must be joined with more active tonics, or the tonics may be exhibited when a remission occurs. If a more active anti-spasmodic and relaxant is required, the following will answer an excellent purpose:

℞.
    Macrotin,
    Gelsemin ........................ aa. grs. V.
    Asclepin ........................... grs. XX.

Triturate well together, and divide into ten powders. Dose, one, twice or thrice a day. As soon as a remission occurs, administer Cerasein in FIVE grain doses once in four hours, and continue until some three or four of the usual periods for the return of the symptoms are past.

Macrotin has been found serviceable in the treatment of neuralgia. The manner of its employment is the same as for the above.

The Macrotin exercises a peculiar and powerfully sanative influence over the functions of the liver, and to this fact are we to look for a solution of its value in many forms of disease. It imparts a healthful impulse to this organ, and powerfully promotes its secretive power. In long standing hepatic derangements, this remedy can scarcely be excelled in efficacy. Hepatic torpor, indigestion, and all their concomitant symptoms are most effectually obviated by the use of the Macrotin. It is not as prompt in its operation as many other remedies, yet it does its work surely. In order to realise its full and true value, the patient should be kept slightly under the constitutional influences of the remedy, as in other cases, until the symptoms yield. In some cases it may be advisable to occasionally exhibit a dose of Podophyllin, Leptandrin, or some other laxative or cathartic, in order to quicken the action of the bowels when tardy, and so obviate the danger of accumulation. When occasion requires the exhibition of laxatives or cathartics, it is better to administer them independent of the Macrotin.

A tendency to prolapsus and other displacements of the uterus may be benefited and cured by the use of the Macrotin. It should be given in small doses, and long continued. We sometimes combine it with other agents, as follows:

℞.
    Macrotin .......................... grs. V.
    Helonin ........................... grs. XV.

Mix, and divide into ten powders. Dose—one, three times per day. If a laxative tonic be indicated, we substitute Hydrastin for the Helonin. In other cases we employ the following pills, which answer an excellent purpose:

℞.
    Macrotin .......................... grs. VI.
    Helonin ........................... grs. XII.
    Leptandrin ........................ grs. XXIV.
    Mucil. Acacia. .................... q. s.

Make a mass and divide into twenty-four pills. Dose—one or two, twice or thrice a day.

Macrotin has been found highly beneficial in the treatment of gonorrhea, gleet, and spermatorrhea, as an auxiliary to other remedies. It is a powerful alterative, and also promotes the action of other alteratives. For gonorrhea or gleet, it may be combined with Stillingin, Irisin, Phytolacin, Rumin, Ampelopsin, Corydalin, or Chimaphilin. The same will be found valuable in secondary syphilis, and in various forms of dermoid disease. For spermatorrhea, the Macrotin may be combined with Lupulin, Gelsemin, Hydrastin, or Cerasein.

We have cured many cases of intermittent fever by first administering a full cathartic dose of Podophyllin, and then exhibiting the following powders during the intermission:

℞.
 Macrotin ........................grs. VI.
 Xanthoxylin ....................grs. XXIV

Mix, and divide into twelve powders. Dose—one, every three or four hours, as the patient can bear. At other times we have combined the Macrotin with Cornin or Hydrastin, Xanthoxylin, etc.

℞.
 Macrotin,.........................grs. V.
 Cornin, ........................... ʒ ss.

Mix, and divide into ten powders. Dose—same as above.
Or,

℞.
 Macrotin ........................grs. V.
 Hydrastin ......................grs. X.
 Xanthoxylin ...................Ɔj.

Mix, and divide into ten powders. Dose and employment same as above. If the patient be troubled with a relaxed condition of the bowels, the Hydrastin will be inadmissable. In that case the Macrotin and Cornin, or Macrotin and Xanthoxylin will answer a better purpose.

Chronic bronchitis, laryngitis, etc., have been greatly relieved by the use of Macrotin. It may be used alone, or in connection with Prunin, Senecin, Asclepin, Leptandrin, etc.

Macrotin is also valuable as an external application in many forms of disease. For this purpose it may be dissolved in strong alcohol. For ordinary use, the following will answer:

℞.
 Macrotin........................... ʒ L
 Alcohol ............................ ℥ IV.

This is applied in rheumatism, lumbago, neuralgia, spina' irritation, indolent swellings, synovitis, indolent ulcers, rheu matic opthalmia, etc. For promoting absorption in synovial effusions, we use the preparation much stronger:

℞.
  Macrotin ............................. ℨ I.
  Strong Alcohol....................... ℨ IV.

Apply night and morning. Over this we usually apply a bandage wetted in cold water and well protected with dry flannel. The Macrotin is powerfully relaxant, hence as soon as the reduction of the enlargement is effected, the Macrotin should be discontinued, and the parts bathed with a tincture of Hydrastin and Myricin in Alcohol:

℞.
  Hydrastin ........................... ʒ ij.
  Myricin ............................. ℨ ss.
  Alcohol ............................. ℨ IV.

Bathe freely.

The tincture of Macrotin is also excellent for contracted joints, and all cold and indolent local indurations or enlargements.

# AMPELOPSIN.

Derived from *Ampelopsis Quinquefolia.*
Nat. Ord.— *Vitaceæ.*
Sex. Syst.—*Pentandria Monogynia.*
Common Names.— *Woodbine, American Ivy, Five-leafed Ivy, Virginian Creeper, Wild Wood Vine, etc.*
Part Used—*Bark and Twigs.*
No. of Principles, *three*, viz., *resin, resinoid,* and *neutral.*
Properties—*Alterative, diuretic, expectorant, anti-syphilitic, astringent and tonic.*
Employment—*Scrofula, cutaneous diseases, bronchitis, hooping cough, asthma, dropsy, syphilis, diarrhea, and rheumatism.*

As an alterative, the Ampelopsin may be relied upon in all cases where remedies of that class are indicated. It does its work kindly, silently, yet surely. The average dose of this remedy is THREE grains, though in some cases the dose may be advantageously increased to TEN grains.

In the treatment of scrofula, the Ampelopsin will be found one of the most reliable alteratives that can be employed. It seems especially adapted to the cure of this complaint, and in connection with such other general treatment as may be indicated, will seldom disappoint expectation. The better plan is to administer it in from TWO to FIVE grain doses, two hours after each meal. All alteratives operate better if taken into the stomach in the absence of food. The Ampelopsin exercises a remarkable influence over the absorbent system, hence will be found valuable in all cases where tuberculous deposits or indurations are suspected. It is, for this reason, a suitable remedy in incipient phthisis. In order to demonstrate its utility in these as in other complaints, it should be used alone, such attention being paid at the same time to the liver, bowels, and skin, as the circumstances of the case may indicate. If other medicines are indicated, they should, by preference, be alternated with the Ampelopsin. If the liver be inactive, or deranged in any manner, an occasional dose of Podophyllin should be administered. If the functions of the skin are tardy or inactive, an alkaline bath should be administered twice or thrice a week. For this purpose carbonate of soda, saleratus, or hard wood ashes may be employed. When the latter can be obtained, we give it the preference.

℞.
>> Hard Wood Ashes..................one gill.
>> Boiling Water.....................one quart.

Infuse five minutes and strain. Apply tepid, sponging the entire surface, and rub well with a dry towel. If the patient is very feeble, from one half to one pint of common spirits may be added to the above. We give preference to New England Rum. None but those who have experienced the utility of the alkaline bath as an auxiliary in the treatment of scrofula, skin diseases, rheumatism, dropsy, etc., can properly appreciate its value.

Although we are a strong advocate for employing organic remedies in their simple forms, alternating with others where change is necessary, yet we may sometimes effect combinations better suited to individual cases. Thus in scrofula, skin diseases, rheumatism, etc., if the liver be inactive and the

## CONCENTRATED MEDICINES PROPER.

bowels constipated, we may combine the Ampelopsin with such other of the concentrated medicines as are known to be good in those affections, and which will afford the desired chologogue and laxative properties. The following for example:

℞.
  Ampelopsin............................ ʒj.
  Leptandrin ........................... ʒss.
  Mucilage Gum Arabic ................. q. s.

Make a mass and divide into thirty pills. Dose—from one to two, three times per day. This combination will be found of most especial service in the above mentioned diseases, and in bronchitis, laryngitis, hepatitis, and in all affections of the glandular system.

For hooping cough and asthma, the Ampelopsin may be rendered more efficient by combining it with Macrotin, Asclepin, or Eupatorin Purpu.

℞.
  Ampelopsin......................... ʒss.
  Macrotin............................grs. IV.

Mix, and divide into sixteen powders. Dose—one, repeated every four or six hours.

℞.
  Ampelopsin,
  Asclepin........................... aa. ℈j.

Mix, and divide into ten powders. Dose—same as above.

℞
  Ampelopsin
  Eupatorin Purpu. ................... aa. ℈j.

Divide into ten powders and exhibit same as above. Either of these formulas may be employed as may seem best adapted to the case in hand.

The Ampelopsin has proved a reliable agent in the cure of dropsy. Although possessing considerable diuretic power, its curative action in this disease does not seem to depend upon that especial property, but upon its power to excite a healthful action in the glandular and absorbent systems, and of promoting depuration. Its influence seems to be expended upon the entire organism, gently stimulating each function to the performance of its duty, without proving evacuant in one direction more than in another. At times, however, it proves actively diuretic. As a general thing it is better to commence the treatment of dropsy by administering a dose of Podophyllin or Jalapin combined with **Cream of Tartar**. Either of the following will answer:

℞.
    Podophyllin..................................grs. ij.
    Bitartrate of Potassa..................... ℈j.

Administer in a spoonful of water at bed time. As soon as the above has operated thoroughly, commence with the Ampelopsin, and exhibit in doses of from FIVE to TEN grains three times per day. The Podophyllin and Cream of Tartar should be repeated occasionally during the course of the treatment.

Or Jalapin **may be substituted** for the Podophyllin, as follows:

℞.
    Jalapin........................................grs. IV.
    Bitartrate of Potassa ................... ℈j.

In some cases we find the three combined to answer a better purpose

℞.
    Podophyllin..............................gr. j.
    Jalapin ....................................grs. ij.
    Bitartrate Potassa...................... ℈j

In other cases it is better to precede the employment of the Ampelopsin with an emetic of Lobelia. For this purpose the Wine Tincture answers an excellent purpose. From TWO to FOUR drachms of the tincture may be given every twenty minutes until free emesis is produced. If there be reason to suspect acidity of the stomach, twenty grains of the super-carbonate of soda should be added to each dose. Or if this caution has been neglected, and the Lobelia is tardy in operating, a teaspoonful of soda dissolved in half a tumbler of warm water should be immediately administered. The Ampelopsin may be employed as above directed, in connection with an occasional hydrogogue cathartic. As soon as the dropsical symptoms are removed, the system must be braced up with tonics in order to prevent a return. Cornin, Hydrastin, Cerasein, Fraserin, or Eupatorin Perfo., either alone or combined with Iron, will answer a good purpose.

Ampelopsin has considerable reputation in the cure of syphilis. It is employed in the same manner as other alteratives. When thought advisable, it may be combined with Stillingin, Irisin, Phytolacin, or Corydalin. As with other alteratives, we deem it better, as a general thing, to use the Ampelopsin alone, and alternate with other remedies. Its use must be persevered in for a length of time, in order to reap its full utility.

The Ampelopsin possesses slightly astringent properties, and has been found serviceable in certain forms of diarrhea. In these complaints, it may be advantageously combined with Leptandrin, Euphorbin, or Juglandin, when the affection proceeds from a deranged action of the liver.

℞.
    Ampelopsin . . . . . . ℈j.
    Leptandrin . . . . . grs. X.

Form a mass with mucilage of gum arabic, and divide into ten pills. Administer one every two hours until the alvine evacuations assume a healthy appearance.

Or,

℞.
    Ampelopsin . . . . . . ʒss.
    Euphorbin . . . . . grs. VI.

Mix, and divide into twelve powders. Dose, same as above.

Or,

℞.
    Ampelopsin,
    Juglandin . . . . . aa. grs. XV.

Mix, and divide into ten powders. Exhibit in the same manner.

If the affection has arisen from cold, the Ampelopsin should be combined with Asclepin. In colliquative diarrhea it should be combined with more powerful astringents, as Geranin, Rhusin, Myricin, Hamamelin, or Trilliin. Thus its range of application may be varied by judiciously combining it with such other agents as may be required to meet special symptoms.

# GERANIN.

Derived from *Geranium Maculatum.*
Nat. Ord.—*Geraniaceæ.*
Sex. Syst.—*Monodelphia Decandria.*
Common Names.—*Cranesbill, Purple Crowfoot, Alum Root, Spotted Geranium,* etc.
Part used.—*The Root.*
No. of Principles, *two,* viz., *resinoid and tannin.*
Properties—*Astringent, styptic, and anti-septic.*
Employment.—*Dysentery, Diarrhea, hemoptysis, hematuria, passive hemorrhages, apthous sore mouth, leucorrhea, gleet, diabetes,* and all affections of the *mucous surfaces.*

GERANIN is justly considered one of the most valuable of the vegetable astringents. In its action, it differs somewhat from astringents generally, in promoting, instead of suppressing the secretive power of the mucous surfaces, and leaving them moist and invigorated in their functions. This remedy has been largely employed in the treatment of dysentery, and with more general success than any other astringent. Its use is admissible in all the different stages, although success will be more certain if the bowels are first relieved of their morbid contents, and the functions of the liver corrected by the use of Podophyllin, Leptandrin, etc. The medium dose of the Geranin is THREE grains. The doses may be repeated every hour, or once in two, four or six hours according to the urgency of the symptoms. When the discharges from the bowels are profuse, the skin hot, dry, and constricted, and the tongue and fauces red, parched and inflamed, the Geranin will answer an admirable purpose in combination with Asclepin.

℞
    Geranin .................................. ℈j.
    Asclepin................................. grs. X.

Mix and divide into ten powders. One of these may be administered every hour. In a short time after commencing the use of the medicine the mucous surfaces will resume their secretive action and become moist, and a gentle moisture appear upon the skin, while the dejections from the bowels will become less frequent and more healthy in appearance. The dose we have named will not be sufficient in some cases, and must be increased to the production of the desired effect. In all forms of bowel complaints attended with spasmodic pains, and when astringents are indicated, the Geranin is advantageously combined with Caulophyllin.

℞
    Geranin,
    Caulophyllin ........................ aa. ℈j.

Mix, and divide into ten powders. Dose—one, to be repeated every hour or two, as may be necessary. This combination will be found excellent for relieving the griping pains so common in these complaints. In diarrhea and dysentery of a bilious character, a more suitable and efficient combination may be effected with the Dioscorein.

℞.
    Geranin................................... ℈j.
    Dioscorein............................. grs. X.

Mix, and divide into ten powders. Dose—same as above.

This prescription is peculiarly useful in cholera morbus and cholera infantum. In the sinking stages of dysentery and similar affections, the Geranin should be combined with stimulants and tonics. The following we have employed quite extensively, and with excellent results.

℞.
 Geranin,
 Xanthoxylin ........................aa. ℈j.

Mix, and divide into ten powders. Exhibit as above directed. This is excellent in the advanced stages of cholera infantum. When tonics are indicated, Cornin, Cerasein, and Fraserin will be found reliable.

In the advanced stages of all diarrheal complaints, and in all cases where there is a tendency to putrescency of the fluids, the Geranin, when indicated, should invariably be combined with Baptisin.

℞
 Geranin ............................. ʒ ss.
 Baptisin ............................ grs. XV.

Mix, and divide into fifteen powders. Give one every two hours. In some cases, it will be necessary to double the quantity of Geranin. No remedy with which we are acquainted is more to be relied upon for correcting the putrefactive tendency than this. In typhoid and other fevers, inflammation of the bowels, etc., this combination will be found exceedingly useful.

Geranin has been found serviceable in checking hemorrhages from the lungs, stomach, bowels, kidneys, and uterus. The usual dose in such cases is FIVE grains, although as much as TEN grains is sometimes given. The doses are repeated every hour until the hemorrhage is arrested, and then at longer intervals. In passive hemorrhages this remedy has proved itself of great utility. In hemorrhage of the bowels, it is sometimes more efficient when administered by enema. From ONE-HALF to ONE drachm may be so administered at a time, and repeated when occasion requires. It may be added to mucilage of slippery elm, starch water, etc. We have known some cases of dysentery to yield readily to this treatment when remedies by the stomach had failed.

Leucorrhea, gleet, and other affections of the mucous surfaces have been benefited and cured by the use of the Geranin. It is both administered internally and applied externally. For external use it is sometimes made into a tincture and then added to water. At other times it is simply added to warm water, in which, however, it is only partly soluble.

Geranin, in connection with suitable diet and tonics, is of great service in the treatment of diabetes. From TWO to FIVE grains may be given three times per day. The bowels should be kept open by the use of small doses of Podophyllin, Leptandrin, or Juglandin.

The diarrhea occurring in the latter stage of phthisis pulmonalis is more readily controlled by the Geranin than any other remedy with which we are acquainted.

The vomiting in cholera has been checked with Geranin when other means failed.

Externally, the Geranin is employed in a variety of affections. The apthous sore mouths of infants is frequently cured by a wash made by adding half a drachm of Geranin to four ounces of warm water. The same is found serviceable in some forms of opthalmia, otorrhea, sore nipples, eruptions of the skin, chafes, etc. An ointment serviceable in the treatment of piles is made as follows:

℞.
    Geranin .................................... ʒj.
    Lard ........................................ ℥j.

Mix. The following is still better:

℞.
    Geranin .................................... ʒj.
    Hydrastin .................................. ʒss.
    Lard ........................................ ℥j.

Mix. Anoint the parts freely several times a day. The same has been found useful in scaly eruptions of the skin.

Dissolved in alcohol, in the proportion of half a drachm to

the ounce, it is an excellent application for toughening the skin when rendered irritable by shaving.

The Geranin will be found one of the best and most reliable astringents in the range of the Materia Medica, but will fail, like all other remedies, when the indications for its employment are mistaken. Thus we would never think of giving it in dysentery and kindred complaints untill the morbid material of the stomach and bowels had first been removed by suitable remedies, and the action of the liver corrected. And if this be done, the neccessity for astringents will be materially lessened. It is bad practice to treat bowel complaints in their primary stages with astringents, and which cannot be to severely reprehended. Assist nature to expel the morbid material which is the direct cause of the inordinate evacuations, then tone up the various functions that have been weakened by excess of action.

# POPULIN.

Derived from *Populus Tremuloides.*
Nat. Ord.—*Salicaceæ.*
Sex. Syst.—*Diœcia Octandria.*
Common Names.—*Upland Poplar, White Poplar, Quaking Aspen, etc.*
Part Used—*The Bark.*
No. of Principles, *two,* viz., *resinoid* and *neutral.*
Properties—*Alterative, tonic, diuretic, stomachic, depurative, vermifuge, and diaphoretic.*
Employment—*Indigestion, flatulence, worms, hysteria, jaundice, fevers, cutaneous diseases, scalding and suppression of urine, night sweats, etc.*

WE shall not, perhaps, have occasion to speak of any remedy more reliable than the Populin in fulfilling certain indications. We have used it long and extensively, and always with the most gratifying results. As a remedy for indigestion accompanied with flatulence and acidity, we know of no single agent more to be relied upon than this. The average dose of the

Populin in these cases is THREE grains three times per day. It will have a better effect if taken immediately after eating. We have found by experience that all medicines calculated to promote digestion and prevent acidity and flatulence answer a much better purpose when administered *at the time* their action is needed. It is presumed that the therapeutic properties of such remedies are, in a measure, expended locally. Hence it is proper to administer them at those periods when such local excitement is necessary. Alteratives, on the contrary, operate better when taken into the stomach in the absence of food, as they are then enabled to be digested, absorbed and conveyed to their destination by the undivided forces of the system.

The dose of the Populin will vary from TWO to SIX grains according to the impressibility of the patient's system, or the effect desired to be produced. In small and oft repeated doses it powerfully promotes diaphoresis. In large doses it proves more actively diuretic. Hence, in the treatment of fevers, it should be given in small quantities and often; while in suppression, retention, and scalding of the urine, the doses should be larger, and exhibited at longer intervals.

For the removal of flatulence it is more of a radical than an immediate remedy, overcoming the disposition by its powers as a corrective. It will be found one of the most certain remedies for this purpose that has ever yet been discovered.

For removal of worms it should be given in from THREE to FIVE grain doses three times per day for a few days, and be followed by a cathartic.

In hysteria it is mainly useful as a tonic after the urgent symptoms are quelled. For this purpose it will be found of singular utility, as it will be tolerated by the stomach when other tonics are rejected, and tranquilise the sympathetic disturbance arising from uterine excitement. It is, for this reason, an excellent remedy for the dyspeptic symptoms accompanying pregnancy.

In jaundice the Populin is of eminent service. It possesses the properties of an alternative to a marked extent, which is displayed by its power to correct the secretive action of the skin and kidneys. It is of great importance that these emunctories should be restored to a normal condition in the treatment of jaundice, as they constitute the main channels of depuration. To render the Populin more effectual, it should be alternated with alterative doses of Podophyllin, Leptandrin or Juglandin.

Populin is one of the most reliable remedies for the relief of night sweats that it has ever been our good fortune to become acquainted with. We refer its curative action in this instance to its power of restoring and giving vigor to the secreting vessels of the skin. This property we have referred to in speaking of the Senecin. For the cure of the above complaint, when not arising from hepatic congestion, FIFTEEN to TWENTY grains of Populin should be administered daily. We usually employ it in solution.

℞
  Populin - - - - - - - Ɔj.
  Warm Water - - - - - ℥iij.
Mix. The Populin is not entirely soluble in water, yet sufficiently so for all practical purposes. It should be stirred up when taken. One tablespoonful of the above solution should be given once in two hours.

Suppression and retention of urine are readily relieved with the Populin, for which purpose it may be used in such doses, and with such frequency of repetition as the case demands. All the directions we deem necessary are, to give it in solution, and in sufficient quantities to produce the desired effect.

Valuable as we deem the Populin in the treatment of the affections previously named, it has one other property which we consider of paramount importance to all the rest, and that is, its property of relieving painful micturation, and heat and scalding of urine. Did it possess no other curative value, we should esteem it an indispensible constituent of our materia medica. Its value in this respect is most apparent when the symptoms above named occur during pregnancy. The relief it affords is

## CONCENTRATED MEDICINES PROPER.

most gratifying to both patient and practitioner. Our method of employing it is in solution, in connection with tincture of Gum Myrrh, as follows:

℞.
    Populin .............................. ℈j.
    Tinc. Myrrh ........................ ʒij.
    Warm Water ...................... ℥iv.

Of this mixture one tablespoonful may be given once every two to four hours, and continued until the symptoms are entirely relieved. In order to allay the irritation of the meatus urinarius and labia, we employ the following:

℞.
    Pul. Gum Myrrh .................... ℥ss
    Boiling Water....................... O.ss.

Infuse and strain. Wash the parts freely with this infusion, or a cloth wetted with it may be inserted between the labia, and in contact with the meatus. This treatment will seldom or never disappoint the practitioner. We look upon it as the most certain prescription that can be made. We can recollect of no instance of failure. It is perfectly safe in all stages of pregnancy.

Many combinations may be effected with the Populin, some of which we are in the habit of dispensing daily. We give below our favorite formulas:

℞.
    Populin
    Xanthoxylin...........................aa. ℥ss.
    Mucil. Acacia ........................q. s.

Form a mass and divide into twenty pills. Or what will answer equally as well, if not better, the following:

℞.
    Populin ............................. ʒj.
    Con. Tinc. Xanthoxylum................q. s.

Form a mass and divide into twenty pills. These pills are serviceable in debility, indigestion, loss of appetite, flatulence, acidity of the stomach, etc. We direct one to be taken immediately after each meal. The stimulant properties of the

Xanthoxylin increase the efficacy of the Populin in cases where great coldness and inactivity of the system exist.

In cases of hepatic torpor and constipation, we employ the annexed formula:

 ℞.
  Populin,
  Leptandrin ........................aa. ℨj.
  Con. Tinc. Xanthoxylum ................q. s.

Form a mass and divide into thirty pills. Use in the same manner as above directed. These we find excellent for promoting the secretions of the liver and obviating constipation. When the difficulty has been of long standing, Phytolacin may be substituted for the Leptandrin.

From the description we have given of the properties and employment of the Populin, the practitioner will be enabled to effect many valuable combinations not necessary for us to notice here. In consequence of the hygroscopic property of the neutral principle of the Populin, it is necessary to make it into pills or reduce it to solution when consecutive doses are prescribed. If preferred it may be dissolved in alcohol, in which it is soluble in equal proportions.

We would earnestly call the attention of practitioners to the Populin, assuring them that they will find it a reliable remedy in fulfilling the indications we have named. It has proved so useful in our hands that we are anxious that all should avail themselves of its valuable remedial properties in the treatment of disease. We trust to the discriminating intelligence of the profession to decide that we have not over-rated its medicinal worth. A fair trial of its merits will confirm the opinion that it is truly a *positive medical agent.*

# CYPRIPEDIN.

---

Derived from *Cypripedium Pubescens.*
Nat. Ord.—*Orchidaceæ.*
Sex. Syst.—*Gynandria Diandria.*
Common Names.—*Wild Ladies Slipper, Yellow Umbel, Nerve Root, American Valerian, Moccasin Flower, etc.*
Part used.—*The Root.*
No. of Principles, *two,* viz., *resinoid and neutral.*
Properties—*Anti-spasmodic, nervine, tonic, and narcotic; also, diaphoretic.*
Employment.—*Hysteria, chorea, nervous headache, neuralgia, hypochondria, nervous irritability, fevers, debility,* etc.


THE Cypripedin fully represents the therapeutic properties of the plant. It is frequently employed as a substitute for the imported valerian, but it will not be found identical with it. As a nervine and anti-spasmodic, the plant has long been used in domestic practice, and with the most beneficial results. Its concentrated equivalent, Cypripedin, possesses the properties above attributed to it in an eminent degree. When opium and its preparations will not agree, the Cypripedin may be relied upon with much confidence. As a substitute for Paregoric, Godfrey's Cordial, etc., it is most advantageously employed in alleviating the disorders of children requiring the use of an anodyne. It possesses, however, some narcotic power, and many times will be found quite as inadmissible as opium. Cypripedin is much used in the treatment of fevers, pleurisy, rheumatism, etc., on account of its anodyne, diaphoretic, and febrifuge properties, It allays pain, abates delirium, promotes perspiration, and procures sleep. It may be given alone in doses of from TWO to FOUR grains, or combined with such other remedies as are being prescribed. In febrile diseases it is employed mostly in combination with Asclepin. The neutral principle of the Cypripedin has a strong affinity for water, and is, therefore, liable to absorb moisture and harden when exposed to the air. For this reason it is necessary to reduce it to solution, or form it into pills, when more than a single dose is to be left with the patient. We employ it mostly in solution.

℞.
  Cypripedin .............................. ℈j.
  Asclepin ................................ ℈ij.
  Warm Water............................. ℥IV.

Dose—from two to four teaspoonfuls once in two hours. As stated under the head of Asclepin, we have seen severe attacks of pleurisy cured with this formula alone.

This formula will be found useful in all febrile diseases attended with nervous irritability. Rheumatism, gout, neuralgia, hysteria, and all spasmodic affections afford indications for its use. In the treatment of scarlatina and other exanthematous fevers, the combination above given will answer an excellent purpose for producing diaphoresis and quieting nervous excitement. Nervous headache is also relieved by administering two teaspoonfuls of the solution every twenty minutes until the violence of the symptoms is abated, then once every hour until complete relief is obtained. A better combination for this purpose may be made as follows:

℞.
    Cypripedin .......................... grs. X.
    Asclepin
    Scutellarin ........................ aa. ℈j

Mix, and divide into ten powers. Dose—one, every twenty or thirty minutes, in warm water. As soon as the symptoms begin to abate, the medicine may be given at longer intervals.

Cypripedin may be joined with Caulophyllin, Lupulin, Viburin, Scutellarin, or other nervines and anti-spasmodics, in the treatment of chorea, hysteria, hypochondria, nervous debility, etc. In many cases it is desirable to combine it with tonics, in which case it may be joined with Cornin, Cerasein, Hydrastin, Euonymin, Fraserin, or Cerasein, accordingly as the properties possessed by either are indicated. All anti-spasmodics are tonics, yet their anti spasmodic power is hightened, or rather confirmed, by joining them with pure tonics. For this reason the Cypripedin, when employed in nervous affections attended with marked periodicity, should be joined with suitable tonics.

As an adjunctive to other remedies, it has been found highly serviceable in dyspepsia, and other affections of the stomach and bowels. It qualifies the action of Cathartics, and abates the tendency to delirium in fevers. Its properties are so well defined, and its uses so generally understood, that we deem it unnecessary to dwell longer upon the manner of its employment.

The practitioner will find it a valuable adjunctive in a great variety of cases, inasmuch as its more prominent properties are so frequently indicated. The large class of diseases to which females are subject afford numerous opportunities for its employment. Although in general agreeing well with the patient, it must be borne in mind that it possesses a degree of narcotic power, and will, therefore, be sometimes found quite as incompatible as opium or any of its preparations. The average dose of the Cypripedin is THREE grains, yet in some cases HALF a grain will be sufficient, while in others TEN grains will be required.

# CHIMAPHILIN.

Derived from *Chimaphila Umbellata*.
Nat. Ord.—*Ericaceæ*.
Sex. Syst.—*Decandria Monogynia*.
Common Names.—*Prince's Pine, Pipsissewa, Wintergreen, Ground Holly*, etc.
Part used.—*The Plant*.
No. of Principles, *three*, viz., *resin, resinoid*, and *neutral*.
Properties.—*Alterative, tonic, diuretic*, and *astringent*.
Employment.—*Scrofula, rheumatism, dropsy, gonorrhea strangury, gravel, debility*, etc.

THIS elegant remedy is now presented for the first time to the profession. The well known efficacy of the plant as an alterative has long rendered it desirable that it should be prepared for medicinal use in a convenient and reliable form. This has been accomplished in the article under consideration. The active principles of the plant, three in number, are here presented, condensed, definite, uniform, and reliable. The average dose of the Chimaphilin is THREE grains. Of course the quantity must be varied to suit the peculiarities of the case in hand. In the treatment of scrofula it will be advisable to administer it in doses of from TWO to FIVE grains three times per day, continuing its use for two or three weeks, and then alternating with some other alterative. Of the latter Ampelopsin, Corydalin, Irisin, Phytolacin, or Stillingin may be selected, as may be best suited to the case. We set a high estimate upon the alterative power of this remedy, an opinion based upon experience. Its operation is not attended with any special excitement, nor is one function apparently stimulated more than another, except it be, in some instances, the kidneys. The whole system seems to be embraced in its influence, manifested by a simultaneous improvement of the various functions of digestion, nutrition, and depuration.

Chronic rheumatism has been frequently relieved and cured by this remedy. As a general thing, larger doses are required than in the preceding case. From FIVE TO TEN grains may be given three times per day. At the same time the bowels should be kept in a soluble condition by the use of Podophyllin, Leptandrin, Juglandin, Euonymin, etc. In this case, as in the former, the Chimaphilin should be alternated with other alteratives, as more satisfactory results will be obtained, as a general thing, by alternation than by combination. Yet there are circumstances and conditions when combinations will meet the indications with greater certainty and promptitude. For instance, in the treatment of rheumatism, ulcers, and other diseases attended with a cold, languid condition of the system, viscidity of the secretions, etc., joined with stimulants, such as the Xanthoxylin, Sanguinarin, or Phytolacin, it will be rendered much more active. Either of the following formulæ may be employed, and will be found excellent:

℞
    Chimaphilin  · · · · ·  ʒss.
    Xanthoxylin · · · · · ·  ℈j.

Mix, and divide into ten powders.
Or,
℞
    Chimaphilin  · · · · · ·  ʒj.
    Sanguinarin  · · · · ·  grs. iij.

Mix, and divide into twelve powders.
Or,

℞.
    Chimaphilin - - - - - - ℈ij.
    Phytolacin - - - - - - grs. X

Mix, and divide into ten powders. One of either of the above powders may be given twice or thrice a day, as circumstances require.

This remedy has been of much utility in the treatment of dropsy, particularly ascites. It seems to act in this complaint much in the same manner as the Ampelopsin, by general and not by specific therapeutic impression. Its value is more apparent in cases originating from or accompanied with an impaired action of the digestive and nutritive system, and debility. In these cases it operates by promoting the appetite, digestion, and assimilation, and gently stimulating absorption and depuration. In the treatment of dropsy, it may be advantageously combined with other of the concentrated medicines suited to the features of the case. Thus, in dropsy of the abdomen, and general anasarca, we should combine it with the Ampelopsin.

℞.
    Chimaphilin,
    Ampelopsin - - - - - aa. ʒ ss.

Mix, and divide into twelve powders. Dose—one, every four to six hours. If more of the stimulant property were needed, we should add a portion of Sanguinarin to the above. The formula would then stand thus:

℞.
    Chimaphilin,
    Ampelopsin - - - - - aa. ℈ij.
    Sanguinarin - - - - - grs. X.

Mix, and divide into twenty powders. Use in the same manner.

In hydrothorax, or dropsy of the chest, we should combine it with Digitalin.

℞
    Chimaphilin - - - - - ℈ij.
    Digitalin - - - - - - grs. ij.

Triturate thoroughly together and divide into ten powders. One of these may be given every five hours, until a perceptible impression is made upon the system in some way, either upon the pulse, kidneys, or respiration, and then at longer intervals, and continued until the symptoms are removed, or there is obvious disagreement of the remedy. In administering this prescription particular care should be taken to neutralise undue acidity of the stomach. As a general thing, it will be better to combine a few grains of super-carbonate of soda with each dose.

Other diuretics, as the Eupatorin Purpu., Lupulin, Populin, Senecin, etc., may be joined with Chimaphilin at the option of the practitioner. For strangury and gravel, we prefer the Populin.

℞
    Chimaphilin .......................... ℨss.
    Populin............................. ♍j.

Mix, and divide into ten powders. Give one every two hours until relief is obtained, then every four or six hours till a cure is effected. The same formula will be found excellent for loss of appetite, indigestion, debility, etc. In these cases one powder may be given twice or thrice a day. Whenever laxatives or cathartics are needed, they should be alternated with the Chimaphilin.

Chimaphilin is very valuable in the treatment of gonorrhea, syphilis, and mercurial diseases. It must be used freely and persevered in for a length of time, occasionally alternating with other tonics and alteratives. When deemed appropriate, it may be joined with Corydalin, Senecin, Irisin, Stillingin, Phytolacin, Rumin, etc., with either of which it is not only admirably suited to the cure of the above affections, but also skin diseases, ulcers, scrofula, and all complaints arising from or accompanied with a vitiated condition of the blood and fluids. As an alterative, and as a remedy in rheumatism, gouty and gravelly affections, chronic cough, and dropsical diseases, it may at all times be relied upon with confidence as an auxiliary, if not as a radical remedy.

# DIOSCOREIN.

Derived from *Dioscorea Villosa.*
Nat. Ord.—*Dioscoreaceæ.*
Sex. Syst.—*Diœcia Hexandria.*
Common Names.— *Wild Yam, Colic Root, etc.*
No. of Principles, *three,* viz., *resin, neutral* and *muci-resin.*
Properties.—*Anti-spasmodic, expectorant, and diaphoretic.*
Employment.—*Bilious colic, cholera morbus, nausea attending pregnancy, spasms, coughs, hepatic disorders, after-pains, flatulence, dysmenorrhea,* and in all cases where an *anti-spasmodic* is required.

THE wonderful efficacy of this remedy in the cure of bilious colic renders it an indispensable agent to every practitioner of the healing art. In this complaint it is as near a specific as any remedy can well be. The relief it affords is both prompt and certain. But its entire value does not relate to this disease alone, as it has been found exceedingly valuable in the complaints above enumerated.

The Dioscorea has been in use, in the crude state, for some considerable time, but we have the pleasure of being the first to record a history of its true concentrated equivalent, Dioscorein, for the benefit of the profession at large. True, a preparation *called* Dioscorein has been offered them, under the designation of a *resinoid*, and represented as being *the* active principle of the plant. By referring to the head of this article, the reader will perceive that the therapeutic properties of the plant reside not in *one*, but in *three* distinct proximate principles, viz., a *resin, neutral, and muci-resin.* The characteristics of these several principles have been described in the first part of this volume. With the exception of the above named *resinoid* Dioscorein mentioned by some authors, the only other method recommended for employing the Dioscorea is in the form of a *decoction.* In this form it has been successfully employed in bilious colic, cholera morbus, etc., proving thereby that it yielded at least sufficient of its properties to *water* to prove actively medicinal. The reader will please remember that *resinoids* are *soluble only in strong alcohol*, hence, if the active properties of the plant had resided in a *resinoid*, the water would have failed in extracting it, and the decoction would be, consequently, useless. But now that we have set the matter in its proper light, there will be no difficulty in perceiving that water may extract a soluble *neutral* and *muci-resin*, and a partially soluble *resin*. We have deemed it necessary to enter thus into detail, in order that the reader might perceive the justice of our charge of inaccuracy against the representation of a *resinoid* being *the* active principle of the plant. We labor for the cause of truth and accuracy in medical science, and we desire that all we write or say shall be capable of demonstration, hence our digression.

The usual dose of the Dioscorein in the treatment of bilious colic is FOUR grains, repeated every thirty minutes until complete relief is obtained. The relief afforded is as prompt as it is certain. In some cases we deem it better to combine the Dioscorein with Asclepin as follows:

℞.
    Dioscorein ............................. ℈j.
    Asclepin................................. ʒss.

Mix and divide into ten powders. Give one every twenty or thirty minutes until the symptoms are fully abated. We have known a single dose of the above to afford entire relief in twenty minutes, rendering further medication unnecessary. In many, however, it will be necessary, in order to effect a radical cure, to follow with a full dose of Podophyllin, which, in cases like this, should be combined with Caulophyllin. The above formula is not only reliable in the treatment of bilious colic, but also in flatulent colic, borborygmus, spasms, etc.

In the treatment of cholera morbus, the Dioscorein should be given in doses of from ONE to TWO grains every twenty minutes until the symptoms are abated. In this case, as in all others, the acidity of the stomach must be neutralised, otherwise the medicine may be of no effect. This may be done by combining a few grains of soda with each dose. In our experience of the management of cholera morbus, as well as of vomiting from other causes, we have found that *small doses frequently repeated* will oftentimes control the symptoms when large doses fail. Hence we deem it expedient in some cases, to give from ONE-FOURTH to ONE-HALF a grain of the Dioscorein at a dose, and repeat every five or ten minutes. The stomach will frequently tolerate and retain very small doses when larger ones are rejected.

We have found the Dioscorein valuable in the treatment of hepatic disorders, particularly when accompanied with irritability of the stomach, and spasm. We generally employ it as an adjunctive to chologogues, as the Leptandrin, Juglandin, etc. Either of the following formulas will be found of excellent service in the treatment of both acute and chronic disorders of the liver.

℞
    Dioscorein ............................... Ɔj.
    Leptandrin............................... Ɔij.
    Mucil. Acacia............................ q. s.

Make a mass, and divide into twenty pills. From one to two of these may be given twice a day.

℞.
    Dioscorein ............................... Ɔj.
    Juglandin............................... ʒj.

Mix, and divide into twenty powders. One of these may be given every four or six hours. The latter will be found excellent in those cases of indigestion accompanied with acidity, flatulence, and spasmodic pains. When the symptoms are aggravated by eating, one of the above powders should be given immediately after each meal. If preferred the powder may be formed into pills with mucilage of gum arabic.

We have found the Dioscorein excellent for allaying the intestinal irritation sometimes produced by Podophyllin. We employ either of the following formulas, accordingly as we wish to secure a diaphoretic or stimulant property.

℞
    Dioscorein........................grs. X.
    Asclepin..........................Ɔj.

Mix, and divide into ten powders. Give one every two or three hours.

Or,

℞
    Dioscorein........................grs. X.
    Xanthoxylin.......................Ɔj.

Mix and divide into ten powders. Dose—same as above. Both of these formulas will be found excellent in diarrhea, dysentery, cholera infantum, etc., at the proper stages.

With Caulophyllin, Viburnin, Scutellarin, Cypripedin, or Lupulin, the Dioscorein is advantageously employed in the treatment of female affections, as hysteria, dysmenorrhea, after-pains, etc. It is an excellent remedy in all spasmodic affections, either as a radical or an auxiliary agent. It may be combined with one or more of the above, as may be best suited to the case. At other times it will require to be combined with tonics, as the Cornin, Cerasein, Fraserin, Hydrastin, Eupatorin Perfo., etc.

Dioscorein has been spoken of as a remedy for the nausea accompanying pregnancy, but we have no personal knowledge of its efficacy in that affection. Judging from its action in other cases, however, we do not hesitate to recommend it for that purpose, confident that if it fails to alleviate, no harm can arise from its administration.

As an expectorant, the Dioscorein has obtained some repute in the cure of asthma, hooping cough, and bronchitis. For asthmatic affections it may be joined with Apocynin, Sanguinarin, Eupatorin Purpu., or Hyoscyamin. For hooping cough, with Macrotin, Asclepin, or Wine Tincture of Lobelia. For bronchitis, with Ampelopsin, Stillingin, Leptandrin, or Prunin.

In conclusion we would reiterate the fact that Dioscorein is eminently anti-spasmodic and diaphoretic, and that its power of relieving spasms relates more particularly to the stomach and bowels, in the disorders of which it has become to be looked upon by many as nearly a specific. We speak of our own knowledge when we state it to be the most reliable remedy yet discovered for bilious and flatulent colic, and intestinal spasm and irritation generally. It is a safe and harmless remedy, but in over doses will produce vomiting.

# CHELONIN.

---

Derived from *Chelone Glabra*.
Nat. Ord.—*Scrophulariaceæ*.
Sex. Syst.—*Didynamia Angiosperma*.
Common Names.—*Balmony, Snake Head, Turtlebloom, Turtlehead, Salt Rheum Weed*, etc.
Part used.—*The Herb*.
No. of Principles, *two*, viz., *resinoid* and *neutral*.
Properties.—*Laxative, tonic*, and *anthelmintic*.
Employment.—*Dyspepsia, jaundice, constipation, debility and worms.*

CHELONIN is of especial value in the treatment of hepatic disorders, and forms a very appropriate adjunctive to other remedies. In the cure of jaundice, it is of eminent service. It seems to stimulate the secretive power of the liver in a peculiar manner, at the same time giving tone and regularity of action. As a tonic, its influence seems to be expended mainly upon the digestive apparatus, increasing the appetite, promoting digestion and assimilation, and so conducing to an improved condition of the blood, both in quality and volume. Being somewhat laxative, it generally obviates constipation. When not sufficiently so, it may be combined or alternated with other laxatives, as the Leptandrin, Hydrastin. Euonymin, etc. The average dose of the Chelonin is THREE grains, yet profitably increased to FIVE or TEN in some cases. In dyspepsia accompanied with hepatic torpor, the Chelonin will be found a most useful agent. The doses may be repeated three or four times a day, as thought necessary. The same is true in relation to jaundice. In the treatment of the latter complaint, a dose of Podophyllin and Leptandrin should be administered once or twice a week.

Combined with Juglandin, the Chelonin will be rendered more efficient in those cases of indigestion accompanied with acidity and flatulence.

℞.
    Chelonin............................Ɔj.
    Juglandin...........................ʒss.

Mix, and divide into ten powders. Dose—one, three times per day. Or with Populin:

℞.
    Chelonin,
    Populin........................aa. ʒj.

Form a mass with mucilage of gum arabic, and divide into thirty pills. Give one immediately after each meal. The same formula will be found excellent for the removal of worms. Two pills may be given three times a day, for three days, and then followed by a cathartic dose of Podophyllin and Leptandrin. If the first trial should prove ineffectual, repeat in the same manner.

We have succeeded in removing large numbers of the *ascaris vermicularis* with the following formula, administered by way of enema, blood warm:

℞.
    Chelonin ........................ ʒss.
    Wine Tinc. Lobelia............... ʒss.
    Warm water...................... ʒIV.

Mix, and administer at once, with a common syringe, and repeat in two hours, if the first dose does not dislodge the vermin. This enema may be repeated every day for a week, or so long as it continues to bring away any worms. We remember several cases permanently relieved by this treatment.

For the removal of the *ascaris lumbricoides*, and *tricocephalus dispar*, the Chelonin may be combined with Gelsemin.

℞
  Chelonin .............................. ℈ij.
  Gelsemin .............................. grs. V.

Mix, and divide into ten powders. Give one three times per day for two or three days, then administer a cathartic.

The Chelonin will be found of excellent service in the convalescing stages of fevers and other acute diseases. It is particularly useful in dysentery after the inflammatory symptoms have subsided, in which complaints it may be combined with astringents, as the Geranin, Myricin, Rhusin, etc., or with diaphoretics, as the Asclepin; or with other tonics, as the Fraserin, Cornin, Cerasein, Populin, according to the particular requirements of the case. It is of especial benefit in all cases where the system has undergone depletion by hemorrhage or colliquitive discharges. When astringents and tonics are indicated, the following is excellent:

℞.
  Chelonin .............................. ℈ij.
  Geranin .............................. ℈j.

Mix and divide into twenty powders. Give one every four hours. When tonics and diaphoretics are needed, we employ the annexed formula:

℞.
  Chelonin .............................. grs. X.
  Asclepin .............................. ℈j.

Mix, and divide into ten powders. Give one every two hours. To enhance the tonic power of the Chelonin in the cases last cited, we prefer the Fraserin:

℞
  Chelonin .............................. ℈j.
  Frasesin .............................. ʒ ss.

Mix, and divide into ten powders. Give one every four or six hours. If it be desirable to increase the tonic and laxative power of the Chelonin, we prefer the Hydrastin:

℞.
  Chelonin............................. ℨss.
  Hydrastin............................ grs. XV.

Mix, and divide into fifteen powders. Give one every four hours. But it must be borne in mind that the Hydrastin will not be admissable in any case where there is acute or subacute gastritis or enteritis, nor in any case of inflammation of the intestinal glands.

# HELONIN.

Derived from *Helonias Dioica.*
Nat. Ord.—*Melancthaceæ.*
Sex. Syst.—*Hexandria Trigynia.*
Common Names. — *False Unicorn, Drooping Starwort, Helonias, Devils Bit, etc.*
Part used.—*The Root.*
No. of Principles, *one*, viz., *a neutral.*
Properties—*Alterative, tonic, diuretic, vermifuge and emmenagogue.*
Employment.—*Prolapsus uteri, amenorrhea, dysmenorrhea, leucorrhea, to prevent miscarriage, dyspepsia, worms. etc.*

No agent of the materia medica better deserves the name of *uterine tonic* than the Helonin. The remarkable success attending its administration in the diseases peculiar to females has rendered it an indispensable remedy to those acquainted with its peculiar virtues. Like the the Senecin, it is alike appropriate in the treatment of diseases apparently calling for dissimilar properties, as, for instance, amenorrhea and menorrhagia. By referring to our remarks under the head of Senecin,.

the reader will there find an explanation of our views upon this subject, and thus save us the necessity of a recapitulation. Its alterative and tonic influence will account, in a measure, for its utility in those complaints. In the treatment of amenorrhea, it will be found most beneficial in those cases arising from or accompanied with a disordered condition of the digestive apparatus, and an anemic habit. It invigorates the appetite, promotes digestion and depuration, and so improves the quality and increases the volume of the blood. In this way the foundation for a cure is laid by improving the tone of the entire system. Aside from this, it has an especial influence over the organs of generation, independent of its general constitutional influence. For this reason it has proved of eminent value in the cure of prolapsus uteri, tendency to miscarriage, and atony of the generative organs. Sterility and impotence have also been relieved and cured by this remedy. In consequence of the peculiar value of Helonin in the treatment of the above named affections, certain writers have classed it as an aphrodisiac, and stated that its continued use induces an abnormal desire for sexual indulgence. Such a statement could only have been made in the absence of actual knowledge, and as the legitimate fruit of a prurient imagination. We have probably used the Helonin quite as extensively as any other practitioner, and we must confess to a want of sufficient penetration to discover any such results from its employment. The only aphrodisiac we recognize, is the natural proclivity of a sensual mind. That the Helonin is a special tonic to the organs of reproduction we are well aware, but only to a normal and healthful extent. Did its action extend further than this, it would be a disease-producing and not a disease-curing remedy. When a medicine so acts upon a diseased organ as to restore it to a physiological condition, we very naturally conclude that said organ will manifest the fact of its restoration by the resumption of its functional activity. This is precisely the case when the Helonin is employed. If administered for the cure of indigestion, the appetite improves, the food is digested, absorbed and assimilated, and thus is the curative action of the remedy manifested. If, on the other hand, the case be one of amenorrhea, sterility, menorrhagia, or impotency, secretion is restored, tone imparted, and the healthful flow of returning stimulus is manifested by the usual physical signs. The sexual appetite is the sequent and not the antecedent of the restoration of the ability of the organs to perform the functions assigned them by nature. Too much confidence must not be placed in the statements of writers who are deficient in clinical experience, and who write only from report, or who assume to know too much, and who, therefore, become ridiculous as well as untruthful.

The Helonin being composed entirely of a neutral principle, is, therefore, mostly soluble in water, in which vehicle it is best administered. For the same reason, as a tonic, it will be tolerated by the stomach when other tonics are rejected. Containing no resinoid principle, it is completely soluble in the stomach, and is, therefore, an appropriate tonic in the convalescing stages of dysentery and other intestinal diseases. Its operation is entirely devoid of irritation.

The average dose of the Helonin is THREE grains, which dose may be repeated three times per day. In the treatment of prolapsus uteri, the organ should first be replaced and quiet enjoined upon the patient, if necessary in the recumbent position, and the Helonin then administered in doses of from TWO to FOUR grains three times per day. The cure may be facilitated by placing a plaster of galbanum, or some other stimulant, upon the sacral region, and the use of the following vaginal enema:

℞.
 Hyarastin........................... ℨss.
 Myricin............................. ℨj.
 Boiling Water.......................O.j.

Infuse and strain. Inject two ounces with a female syringe two or three times a day. If the affection be accompanied with inflammation and slight ulceration of either the os uteri or vaginal walls, we prefer the following:

℞.
    Chloride of Lime . . . . . ℥j.
    Cold Water . . . . . O.ij.

Put the lime in a bottle, add the water, shake well, stand it aside to settle, and use the clear solution in the same manner as the above. If a more stimulating injection seems necessary, add two ounces of the chloride to a quart of water. If the liver is inactive and the bowels inclined to constipation, we combine the Helonin with Leptandrin:

℞.
    Helonin . . . . . . ℈ij.
    Leptandrin . . . . . ℈j.

Form a mass with Mucilage of Gum Arabic and divide into twenty pills. Give one three times per day. Or the Leptandrin may be alternated with the Helonin, two or three grains of which may be administered at bed time. The same plan of treatment will be found equally useful in the treatment of some forms of leucorrhea, particularly those cases accompanied with or arising from prolapsed uterus, debility, etc.

Either alone, or combined with other appropriate remedies, the Helonin will be found reliable in the radical cure of amenorrhea. In simple uncomplicated amenorrhea, it is best joined with Senecin:

℞.
    Helonin,
    Senecin . . . . . aa. ℈j.

Mix, divide into ten powders, and give one three times per day. The same formula will serve an excellent purpose for the cure of dysmenorrhea and menorrhagia, in which complaints it should be administered regularly during the intermenstrual period. Upon the approach of the menstrual molimen its use should be discontinued, and the patient placed under the influence of Caulophyllin, Gelsemin, Viburnin or other anti-spasmodics in dysmenorrhea, and Trilliin, Oil of Erigeron, Lycopin, Geranin or Myricin in menorrhagia. When the period has passed, the remedy should be again resumed.

In anemic habits the Helonin is advantageously joined

with Iron. If hysteric symptoms are present, with the Valerianate:

℞.
 Helonin............................... ℈ij.
 Valerianate of Iron..................... ℈j.

Mix, and divide into twenty powders. Dose,—one three times per day. In defective menstruation we employ the following, which we prefer to any other combination we have ever employed:

℞.
 Helonin ........................... ℈ij.
 Iron by Hydrogen.................. grs. XVI.

Mix, and divide into sixteen powders. Give one morning and evening. If the patient be advanced in years, and irritability of the stomach does not contra-indicate, the Phosphate of Iron may be substituted.

Helonin has been found serviceable in correcting a tendency to miscarriage, which it effects by virtue of its properties as a special uterine as well as a general tonic. In those cases the doses, frequency of repetition, and continuance must be such as the judgment of the practitioner may indicate.

In the treatment of the various forms of dropsy, the Helonin has proved of remarkable utility. It operates in a general manner, and is, seemingly, a powerful resolvent. It restores the appetite, improves digestion, promotes absorption and depuration, and imparts a healthful impetus to the whole economy. The only manner in which it proves visibly evacuant, is, in some cases, as a diuretic, except when given in over doses, in which case it proves emetic. In the treatment of dropsy, it may be combined with Ampelopsin, or Apocynin, or Digitalin, or Sanguinarin, etc. For general anasarca, with Ampelopsin:

℞.
 Helonin............................. ʒss.
 Ampelopsin......................... ʒj.

Mix, and divide into fifteen powders. Give one every four

Twice a week give the following cathartic:

℞.
  Podophyllin............................grs. ij.
  Cream of Tartar......................℈j.

Mix. Administer in a little water at bed time. For dropsy of the abdomen, it may be appropriately joined with Apocynin.

℞.
  Helonin ................................ ʒ ss.
  Apocynin ..............................grs. X.

Mix, and divide into ten powders. Give one three times per day. For hydrothorax, hydrops uteri, and ovarian dropsy, it may be combined with Digitalin.

℞.
  Helonin ................................ ʒ ss.
  Digitalin..............................grs, ij.

Triturate well together and divide into ten powders. Give one, two or three times a day. Be particular to neutralise undue acidity of the stomach previous to the administration of this remedy, and employ a fluid menstruum in exhibiting it. In dropsy of the ovaries the following Liniment will be found a valuable auxiliary:

℞.
  Con. Tinc. Digitalis ..................... ʒ ij.
  Tincture of Squills,
  Alcohol ..............................aa. ℥ ij.

Mix. Bathe the parts freely two or three times a day, or apply a cloth wetted with the liniment. This application powerfully promotes absorption.

When great languor, coldness and debility exists, the Helonin is beneficially joined with Sanguinarin.

℞.
  Helonin................................℈ij.
  Sanguinarin ..........................grs. X.

Mix, and divide into twenty powders. Give one three times per day. To render the prescription more stimulating, Xanthoxylin may be added, as follows:

℞.
Helonin,
Xanthoxylin........................aa. ℈ij.
Sanguinarin............................grs. vij.

Mix, and divide into twenty powders. Dose, same as above. In this way combinations may be effected to suit the peculiarities of the case in hand.

For the removal of worms, the Helonin may be given in FOUR grain doses morning and evening, for two or three days, followed by a cathartic. After the worms are expelled, the Helonin should be continued in TWO grain doses for a time, in order to strengthen the stomach and bowels, and so obviate the condition giving rise to the generation of the vermin.

As a general tonic, in the convalescing stages of fevers, dysentery, and other acute diseases, dyspepsia, etc., the Helonin may at all times be relied upon with much confidence. As a general thing, it should be employed alone when it is desirable to realise its specific influences, yet appropriate combinations may be effected when the practitioner deems it advisable. We have found it useful when joined with Cornin in certain forms of dyspepsia, and with Cerasein in passive hemorrhage and menorrhagia. With Fraserin, it will be appropriate when the system has been exhausted by colliquitive discharges.

As a tonic in debility of the uterus and appendages, we know of no organic remedy deserving of greater confidence. We have used it long and extensively, and with the happiest results. We sometimes join it with Caulophyllin in amenorrhea, and with Baptisin in defective menstruation, and when tonics and antiseptics are indicated, as in typhoid, typhus, and other fevers, dysentery, scarlatina maligna, etc. The dose will vary in different cases, and under different circumstances. We have given the quantity we usually employ in our practice.

# LEPTANDRIN.

Derived from *Leptandra Virginica.*
Nat. Ord.—*Salicaceæ.*
Sex. Syst.—*Didynamia Gymnosperma.*
Common Names.—*Culver's Root, Culver's Physic, Black Root, Tall Speedwell,* etc.
Part Used—*The Root.*
No. of Principles, *four,* viz., *resin, resinoid, alkaloid* and *neutral.*
Properties—*Alterative, deobstruent, chologogue, laxative* and *tonic.*
Employment—*Fevers of every type, dysentery, diarrhea, cholera infantum, dyspepsia, jaundice, piles, laryngitis, bronchitis,* etc.

No one of the concentrated medicines has been so much misunderstood as the Leptandrin. The reason for this resides in the fact that the profession had had but little clinical experience in the use of the plant from which this remedy is derived. Previous to the time of the concentration of the active principles of the Leptandra, little knowledge was to be gained from the various works on materia medica in relation to the therapeutic properties of this plant. The same stereotyped statement was copied into the various publications treating upon therapeutics, the authors seeming to possess little positive knowledge of its virtues, relying rather upon the traditionary reports handed down by the elder botanists. The plant was said, by them, to be possessed of active cathartic properties, and was highly recommended in the treatment of typhoid fever, as it was said to be capable of producing "copious, dark, tar-like dejections from the bowels," and so break up the disease. As soon as the concentrated preparation, Leptandrin, was brought to the notice of the profession, many practitioners commenced employing it in their practice, a large number of whom never had any experience in the use of the plant. Relying upon the truthfulness of the statements they had read concerning the Leptandra, they very naturally supposed that the Leptandrin, being the concentrated equivalent of the plant, was, as there represented, a cathartic of considerable power. Failing to realise such a result from the employment of the Leptandrin, many were disposed to condemn the remedy as being improperly prepared and worthless. Taking advantage of this circumstance, some two or three ignorant and malicious scribblers made themselves not only notorious but ridiculous by attempting to impeach the character of those engaged in the manufacture of the Leptandrin, charging them with fraud and adulteration. But their transparent hypocrisy served but illy to mask the real motives of their canting pretensions. Professing to regard solely the interests of the profession, and to be actuated by a desire to have the profession furnished with pure and reliable remedies, they unwittingly displayed the "cloven foot" of ignorance and personal malice, demonstrating the fact, by their disgraceful failure, that they had but "stolen the livery of Heaven to serve the devil in." We highly approve and honor capable and honest criticism, believing it to be the great conservator of medical science; but we equally deprecate the unworthy attempts of incompetent meddlers with subjects they cannot comprehend, and which they essay only to give vent to the cankering venom so prone to generate in base and ignoble minds. All attempts at imposture in pharmaceutical preparations should be denounced by the unanimous voice of the profession; but even-handed justice demands that such denunciation should follow, and not precede conviction.

In the early history of the Leptandrin, a "resinous substance" was supposed to embody the active properties of the plant, which idea is still indulged by some manufacturers, consequently the preparations they offer to the profession under the appellation of Leptandrin consists mostly of the resinoid principle of the plant, to the exclusion of *three* other important principles, namely, a *resin, neutral,* and *alkaloid.* The article of Leptandrin now under consideration consists of *four* distinct principles, namely, a *resin, resinoid, alkaloid,* and *neutral.* With the assistance of the explanations given in the first part of this volume, any competent chemist may ascertain the truth of our statement by analysis. When this fact was first announced to the profession, accompanied with proof in the form of the article in question, ignorant and interested persons endeavored to cast suspicion upon the character of the preparation by denying the fact of multiplicity of principles, accounting for the obvious difference in its composition, when compared with the "resinous" Leptandrin, by the charge of adulteration and foreign admixture. But unfortunately for the success of charlatans, the science of organic chemistry is sufficiently definite in its manipulations to enable the honest searcher after truth to test the accuracy of all pretensions submitted to its ordeal. Through this ordeal the Leptandrin under consideration has passed again and again, and yet will pass, and thus the claims of truth be vindicated. We desire no one to take our *ipse dixit,* but, if dissatisfied, to boldly, manfully, and independently investigate all matters where contrariety of sentiment is held or expressed. Of such of our readers as are not conversant with the circumstances that have led us into this digression, we humbly beg pardon for taxing their patience with the foregoing preamble; but to those who recognise the application of our remarks, we deem no apology due. If a portion of those engaged in the isolation of the active proximate principles of plants should find themselves less competent and sucessful than others engaged in the same pursuit, let them not seek to divert attention from their own errors and blunders by detraction and defamation, but let them labor rather to correct their own mistakes and defects, and deserve confidence and support by bringing their preparations up to the standard required by the present advanced condition of organic chemical science.

The writer was accustomed, over twenty years ago, to gather and prepare the Leptandra for medicinal use in his father's practice. Many opportunities were then offered for observing its action upon the system. Since that time we have employed the crude powdered root in practice and upon our own person, and have never deemed it more than laxative. It required to be given in repeated doses, at intervals of two hours, in order to obtain an action of the bowels. Its operation would frequently be attended with considerable nausea, griping, drowsiness, and general relaxation of the system. In consequence of the above mentioned symptom of drowsiness having been observed during its operation, some writers have supposed it to be narcotic; but such we do not deem it. We are of the opinion that the symptom arose from the slowness with which the medicine operated, in consequence of the digestive action required to eliminate the active principles from their combination with woody and other inert matters, and partially in consequence of the gradual secretion of morbid matters into the intestinal canal. Be the cause what it might, we never deemed the Leptandra cathartic, although we do not doubt that some practitioners have been deceived into so supposing it in consequence of having administered the remedy at that very moment when nature was ripe for a spontaneous dejection of accumulated fecal material.

Leptandrin is, in our opinion, the most valuable remedy of its class. It is eminently chologogue, resolvent, laxative and tonic. It is slow, but mild, certain, and radical in its operation. No remedy with which we are acquainted is more to be relied upon in chronic affections of the mucous surfaces. Its value in this respect is peculiarly apparent in chronic dysentery and diarrhea, and other diseases of the bowels. When false membranous formations have occurred in the small intestines, produced by the gradual exudation of plastic lymph, the Leptandrin may be relied upon for their removal, with great confidence. The dose of the Leptandrin in such cases will be from TWO to FOUR grains twice or thrice a day, according to the solubility of the bowels. In order to reap its full utility, the remedy must be persevered in for a considerable length of time. Although the Leptandrin may be relied upon alone, we may sometimes effect combinations calculated to accomplish the same object, which, although they may present no apparent advantages, experience has demonstrated to be reliable. The following is with us a favorite formula:

℞
    Leptandrin,
    Juglandin............................aa. ʒj.

Form a mass with mucilage of gum arabic, and divide into thirty pills. Dose—one, two or three times per day. In the treatment of the complaints above mentioned, we have derived the most beneficial results from the employment of the above prescription. We have also used it with great success in the cure of constipation and piles. We recently treated a case of the latter complaint, accompanied with frequent hemorrhage from the rectum, of twelve years standing. A short time after commencing the use of the above remedy, the patient discharged considerable quantities of false membrane in shreds and patches, and a number of pieces several inches in length, forming complete tubes. The evacuation of this matter was attended with an amelioration of all the symptoms, and at the present time the patient declares himself well. The bowels are regular, appetite good, the hemorrhage has ceased, and the distressing pain so long experienced beneath the sacrum entirely gone. We might mention numerous other cases, but it will be of more interest to practitioners to know how to cure their own cases, than to read of those that have been cured.

Leptandrin has obtained a well merited celebrity in the treatment of typhoid and other fevers. Its employment is admissible when more irritating remedies would be objectionable. In typhoid fever, and in dysentery, its action seems to be peculiar and specific. It not only regulates the functions of the liver, but also corrects and restores the secreting power throughout the whole extent of the alimentary canal. Not only does the mucous membrane of the stomach and bowels come under its especial control, but the entire organism acknowledges its sanative power. The whole glandular system, including the skin, partakes of its healthful impress. When the patient is fairly brought under the constitutional influence of the Leptandrin, the skin, which was before hot, dry, and constricted, becomes soft, moist and flexible; expectoration becomes easy, the arterial excitement is lessened, and the patient, before restless, wakeful and delirious, becomes calm, rational, and inclined to sleep. Such are the general constitutional influences of the Leptandrin when administered in acute diseases. In the treatment of typhoid fevers, when chologogues and laxatives are indicated, the Leptandrin should be administered in average doses of THREE grains, every two hours, until sufficient action is produced. One great advantage possessed by the Leptandrin is its tonic power. It never debilitates, but, on the contrary, invigorates while it deterges. The evacuations produced by Leptandrin always give evidence of a sanative influence having been exercised over the secretive functions. In mild cases of dysentery, diarrhea, and cholera infantum, a few grains of Leptandrin will, if administered early, bring about well assimilated fecal discharges in a few hours. In the treatment of all febrile complaints, the Leptandrin is judiciously combined with Asclepin, as follows:

℞.
   Leptandrin,
   Asclepin........................aa. grs. ij.
Mix, and administer at one dose. Repeat once in two hours

until the alvine discharges assume a healthy appearance. These directions apply equally in case of typhoid fever, dysentery, diarrhea, cholera infantum, or other intestinal disorders, Of course, it is expected that practitioners will vary the combination, dose, repetition and continuance according to the necessities of the case.

Leptandrin is one of the best adjunctives to the Podophyllin in all cases when the latter remedy is indicated. We seldom treat either typhoid fever or dysentery, in this locality, without a combination of the two. The following is our usual formula for typhoid fever:

℞.
    Leptandrin ........................grs. iij.
    Asclepin ..........................grs. ij.
    Podophyllin ......................grs. j.

Mix, and give at a single dose. We generally repeat this powder once in twenty-four hours until the secretions of the liver and bowels are corrected. In the treatment of dysentery we employ the following:

℞.
    Leptandrin ........................grs. VI.
    Podophyllin ......................grs. ij.
    Asclepin ..........................grs IV.

Mix and divide into four powders. Give one every two hours, and continue until the discharges from the bowels assume a healthier appearance. We sometimes substitute Caulophyllin for the Asclepin, and we find it excellent for controling the spasmodic pains accompanying this complaint. After the acute symptoms of the disease have subsided, and the bowels continue relaxed, the Leptandrin may be combined with Geranin, Myricin, Rhusin, or other astringents. In this manner the action of each may be modified, and the discharges restrained without producing constipation. As a general thing, however, we prefer to alternate the Leptandrin with astringents, and this plan, we think, will give the practitioner the greatest amount of satisfaction. In all intestinal disorders connected with, or originating from a deranged action of the liver, the Leptandrin is one of the most efficient remedies known. But we would here state, as the result of experience, that when the patient is laboring under obstinate constipation of the bowels, and a cold, inactive condition of the system generally, the use of the Leptandrin should be preceded by a full cathartic dose of Podophyllin, as by so doing, greater promptitude of relief will be ensured. And when the Leptandrin is used as a resolvent and detergent, an occasional dose of Podophyllin or some other cathartic should be administered, otherwise the bowels are liable to become loaded with accumulations of morbid secretions, which if retained, give rise to serious constitutional disturbance. Not only this, but if the bowels move under the influence of the Leptandrin, its operation is generally slow, and the acrid secretions passing off tardily, give rise to a great amount of irritation which, by the above observance, may be avoided. When the Leptandrin is exhibited in small and repeated doses as an alterative, its laxative power becomes considerably modified, hence the necessity of occasionally alternating with a more decided evacuant.

In the treatment of dyspepsia dependent upon hepatic derangement, the Leptandrin will be found one of the most reliable auxiliaries. The same is true in relation to jaundice. It acts in a general and not in a specific manner. It soothes irritability, removes obstructions, promotes secretion and depuration, and imparts tone and vigor of action to the various functions. We have already spoken of its value in the treatment of piles, in which complaint, either with or without hemorrhage, we deem it invaluable. In this affection we generally use it in connection with Hydrastin. They may be combined, or used alternately. If desirable to avoid complexity of prescription, we give the following pill:

℞
    Leptandrin ........................ ℥j.
    Hydrastin ......................... ℥ss.
Form a mass with mucilage of gum arabic and divide into thirty pills. Dose—from one to two, three times per day. At

the same time, if there be hemorrhage or ulceration of the rectum present, we employ the following enema:

℞.
    Hydrastin............................... ʒj.
    Boiling water........................... O. j.

Administer two ounces of the above infusion three or four times per day. Use cold or tepid, as best accords with the feelings of the patient.

As a general thing we prefer to administer from ONE to TWO grains of Hydrastin three times per day, and from TWO to FOUR grains of Leptandrin at bed time. At the same time employ the above enema. Or the combinations of Leptandrin and Juglandin, previously mentioned, may be employed in connection with Hydrastin. One or two of the pills may be given at bed time, and TWO grains of Hydrastin morning and evening. This treatment, if persevered in, will seldom fail of effecting a cure. It is not only necessary to continue the medicine until the immediate symptoms are relieved, but for a considerable time afterwards, in order to strengthen the system against a return. For the latter purpose, the Leptandrin will answer an equally good purpose alone.

Leptandrin has been found very serviceable in the removal of worms. It is usually given in doses of from TWO to FIVE grains twice a day, or in sufficient quantity to keep the bowels somewhat relaxed. It may be advantageously combined with other vermifuge remedies, as Chelonin, Gelsemin, Helonin, Populin, etc. Although sometimes instrumental in expelling worms, its greatest value resides in its power of correcting the action, and giving tone to the bowels after the worms are removed, and so obviating the condition favorable to their generation. For the latter purpose, it may be combined with tonics.

Leptandrin is an admirable auxiliary remedy in the treatment of bronchitis, laryngitis, and other affections of the respiratory organs. It is a safe and certain resolvent, acting in an especial manner upon the mucous membranes, hence is of service in all affections of those surfaces. In chronic inflammation of the bladder, leucorrhea, chronic diarrhea, and dysentery, etc., the practitioner will find it a serviceable and reliable remedy. In the treatment of diseases of the skin, no better general remedy can be brought to bear.

A great and important fact, in connection with the Leptandrin is, that while it promotes and corrects the secreting power of the liver, resolves, deterges, and promotes depuration, it does not debilitate. On the contrary, it is decidedly tonic, giving tone and vigor of action to the entire secretive apparatus of the system. Hence it is always a safe remedy in debility, and in the treatment of the diseases incident to delicate females and infants. For constipation during pregnancy, or for the cure of diarrhea and dysentery, under the same circumstances, and for the intestinal disorders of infants, it is always safe and reliable.

The neutral principle of the Leptandrin is eminently hygroscopic, absorbing moisture from the atmosphere with great readiness, and hardening into a solid mass. For this reason it is inconvenient of dispensation in the form of powder. Where great exactitude is required, it should be formed into pills, or dissolved in alcohol. In the treatment of chronic disease, used either alone or in combination, we frequently deliver it to the patient in bulk, in a well corked vial, directing the proper dose by weight or measure, as by means of a three or five cent piece. Unlike some of the more potent remedies, a slight deviation from the exact quantity will entail no serious consequences. The Leptandrin is neither soluble nor mixable in water, another good reason for forming it into pills. It will mix well with mucilages, as of slippery elm, gum arabic, etc. Average dose, THREE grains.

# DIGITALIN.

Derived from *Digitalis Purpurea*.
Nat. Ord.—*Scrophulariaceæ*.
Sex. Syst.—*Didynamia Angiosperma*.
Common Name.—*Foxglove*.
Part Used—*The Leaves*.
No. of Principles, *four*, viz., *resinoid, alkaloid* and *two neutrals*.

Properties—*Narcotic, arterial sedative, alterative, resolvent, diuretic, antiseptic, etc.*

Employment — *Dropsies, pneumonia,* both acute and chronic, *hemoptysis, neuralgia, mania, epilepsy, pertussis, asthma, rheumatism, disease of the heart,* both functional and organic, *croup, nervous affections* of almost every type, to *prevent abortion, glandular diseases, fever and inflammations generally.* Also in *scrofulous affections, chronic exanthema, local œdema, ulcers, tumors, diseases of the bones and joints, etc.*

The deficiencies of crude organic remedies and so-called officinal preparations have never been more seriously felt, than in the employment of the Digitalis and other plants possessing a high concentration of therapeutic power. The variable amount of active principles residing in the plant has hitherto rendered the employment of the Digitalis somewhat hazardous, as the discrepancies of the plant have attached to all its pharmaceutical preparations. Not only has the *amount* of the medicinal principles present been extremely indefinite, but also the *number*, as the therapeutic properties of the Digitalis reside, not in *one* distinct principle, but in *four*, each one of which represents a more or less distinct medicinal power, and these four, when combined, embody the entire therapeutic value of the plant. Those properties which exercise a peculiar sedative or depressing power over the arterial system, reside chiefly in the resin and oleo-resin; while the neutral and alkaloid principles expend their influence more particularly upon the absorbent vessels. These facts not having been understood heretofore, will account for the many failures and bad results attendant upon the employment of the Digitalis, both in the use of the plant in substance, or of the various pharmaceutical preparations hitherto employed. The plant being uncertain and variable in the actual amount of proximate active principles present, it follows, as a matter of course, that ordinary tinctures, infusions, etc., must, of necessity, partake of the uncertain character of the plant. No process short of isolation and recombination of the various active principles could render the therapeutic powers of the plant uniform, definite, or certain. The Digitalin of which we now propose to treat, is so prepared.

When Digitalin is administered in small and repeated doses to a healthy person, the following symptoms will be developed in the course of from twenty-four to forty-eight hours:—in a majority of cases the secretion of urine will be augmented, and in all cases the secretions of the mucous membranes will be increased; digestion is soon more or less impaired, accompanied with nausea, pain in the stomach, loss of appetite, and colicky pains in the bowels. The effects of the Digitalin are next displayed upon the arterial and nervous systems, the frequency of the pulse is greatly diminished, often being reduced to one half the usual number of beats per minute, and generally becoming small, soft, and feeble. The latter effect, however, only appears after the Digitalin has been exhibited for two or three days consecutively, and usually continues for several days after the use of the Digitalin has been abandoned. In many cases, however, the effect of the Digitalin upon the arterial system is quite the contrary, increasing instead of diminishing the frequency of the pulse, and giving rise to local congestions, hemorrhage of the lungs, etc. It is only in cases of debility that the depressing power of the Digitalin is uniformly and surely manifested upon the arterial system.

When administered in larger doses, the Digitalin first stimulates the arterial system, and gives rise to vomiting, diarrhea, obscured vision, sparklings before the eyes, dilation of the pupil, vertigo, stupor, violent headache, and congestion, etc. But these evidences of irritation do not continue long, soon giving place to symptoms of great depression and paralytic debility. The pulse sinks rapidly, becoming small and unfrequent, followed by great lassitude, faintness, drowsiness, etc., which state frequently continues for several days.

When given in very large doses, the Digitalin acts upon the stomach and intestines much like a caustic poison, producing a severe burning sensation in the throat and stomach, salivation, thirst, spasm of the glottis, painful retching and vomiting of greenish matter, diarrhea, delirium, and convulsions. These symptoms are soon succeeded by insensibility, general paralysis, accompanied with a small, feeble, unfrequent, and often intermittent pulse. This condition, even after the exhibition of moderate doses of the Digitalin, frequently ends in a fatal apoplexy. Upon dissection, when death has ensued, we find the mucous surfaces of the stomach and bowels inflamed and broken down, but seldom is the vascular structure of the head, or the venous system generally, in a congested condition. The lungs usually present a normal appearance.

It has been seen that the Digitalin possesses two primary and distinct therapeutic powers, which expend their influences in different directions. The first exercises a remarkable influence over the heart and arterial system, depressing and retarding their functional activity, while the second property is expended upon the absorbent and venous systems, and upon the lymphatic vessels and glandular structure generally, stimulating them to increased activity. This is the case even when applied externally; as, for instance, when applied to tumors and enlarged glands.

*Digitalin depresses and retards the activity of the* POSITIVE VITAL FORCES *engaged in the processes of organic formation and reproduction; while it stimulates and quickens the activity of the* NEGATIVE FORCES. This fact will be apparent when it is considered that the *arterial* system superintends the conveying of the plastic formative materials of the blood to their proper destinations; while, on the other hand, the *venous* and *lymphatic* systems perform the duty of conveying away, not only the superfluous materials and effete matters given off during the processes of organic formation, but also have to re-dissolve and absorb what has been already formed, particularly when morbidly active, all of which processes are necessary to the institution and completion of the phenomena of reproduction.

A difference of opinion exists as to whether Digitalin acts primarily upon the heart and arterial system in the production of the phenomenon of sedation, or whether this result is the consequence of counter stimulation, and therefore secondary. For our own part, we incline to the former opinion, drawing our conclusions from observations made at the bedside, the only proper place to decide the precise therapeutic operation of remedial agents. We find that the Digitalin, in most cases of an abnormally increased activity of the heart and arteries, relaxes the tone of the arterial vessels, and depresses the action of the heart, diminishes the force and frequency of the pulse, and renders it soft, small, and infrequent. We find, further, that the Digitalin is a most excellent remedy for the relief and cure of those sequela which remain when inflammatory affections have been subjected to the antiphlogistic treatment, manifested by a morbid activity of the whole arterial system, or by some of its single branches. At the same time, its influence over the absorbent vessels promotes the resolution of local inflammations and congestions. Digitalin is, in general, a powerful relaxant and sedative remedy for the relief of a morbidly increased activity of the arterial system, yet, in certain conditions, it will prove a powerful stimulant to the same organs.

No less important is the therapeutic effect produced by Digitalin upon the absorbent system. Its influence is evidently that of a stimulating tonic, and its impressions are not confined to the absorbent vessels, but extend to the veins, glands, mucous, fibrous, and serous membranes, and to the epidermis.

Digitalin is eminently resolvent and alterative, overcoming viscidity of the secretions, and quickening the activity of the entire absorbent system. It excites, in an especial manner, the absorption of serous effusions, and promotes their depuration through the natural channels. From the fact of its influence in increasing the secretive action of the kidneys, it is termed a diuretic. The diuretic effect of Digitalin, however, is not primary, like that of oil of turpentine, cantharides, etc., which operate by direct irritation and stimulation of the urinary organs, but is manifested only in proportion to the degree of absorption excited. Even when Digitalin is given in excess, we do not observe those symptoms of local irritation of the urinary apparatus which attend the administration of the above-named specific diuretics.

In diseases requiring large doses, or the continued use of Digitalin, it will be necessary to counteract the disturbance it usually creates in the functions of digestion and nutrition, as well as the narcotic properties above referred to, and which often render its use objectionable, by the use of suitable remedies. Of the narcotic properties of the Digitalin, we can seldom make any specific use. Thus much of its **therapeutic history**.

EMPLOYMENT.—Among the indications in which Digitalin is employed, we may first mention those conditions characterised by a morbidly increased activity of the arterial system, either throughout its whole extent, or of some of its numerous branches. This condition is manifested more by a quickened pulsation than by an increase of tone. This abnormal excitement of the arterial system may arise from two distinct and separate exciting causes; in the first place, from a superabundance of the materials of excitement in the blood; and, in the second place, from an exalted or morbid irritability of the heart and arterial vessels. It is in either of the above conditions that Digitalin is most successfully employed.

But in many cases it will be found that both causes are operating at the same time, in which event it becomes necessary to combine the Digitalin with other remedies. Under these circumstances the Veratrin is particularly indicated.

The morbid irritability inherent to the heart and arterial system may be produced or aggravated by the continued incitement of reflex action originating from an abnormal condition of the heart itself, of the arteries, lungs, etc.; as organic disease of the heart, ossification of the aorta, tuberculous deposits in the lungs, or organic disease of some other important corresponding organ. In these cases the Digitalin will be found a valuable palliative.

On account of the peculiar influence it exercises over the absorbent system, Digitalin is beneficially employed in the treatment of those diseases arising from or dependent upon inactivity of the lymphatic vessels and glands, serous membranes, and veins, and when it is necessary to stimulate the absorbent functions to increased activity in order to depurate through the urinary canals fluids already secreted or exudated. But when the inactivity is the result of vital exhaustion and debility of the absorbent system, Digitalin is contra-indicated, and its employment will be attended with bad results. *Digitalin may awaken and incite to action the latent or sleeping forces of the system, but it is incapable of infusing vitality or recruiting exhaustion.*

In acute fevers, Digitalin is generally an uncertain and critical remedy, quite frequently producing contrary effects from those desired. It had better, therefore, be avoided in such cases, unless it is clearly and distinctly indicated.

The morbid irritability of the heart and arterial system mentioned above, is often apparent in intermittent and remittent fevers, manifested by an increased action of the pulse, while the temperature of the surface and the rest of the febrile symptoms are not present in a corresponding degree. This exalted sensibility supports and perpetuates the febrile condition, and gives rise to various disturbances of the circulation, such as congestions, etc. Under such circumstances a judicious use of the Digitalin will be attended with beneficial results.

In rheumatic fevers, the Digitalin will not only diminish the fever, but also moderate the profuse symptomatic sweats which attend the disease, and which arise from excessive capillary congestion.

In acute exanthematous fevers, Digitalin is of great value, partly because of the great irritability of the arterial system, and partly because of the great tendency in these complaints to exudation, concretions, etc., and the liability to malignant sequela, which the depurative power of the Digitalin is calculated to obviate.

In lingering hectic and pneumonic fevers, the Digitalin is of much advantage, either when the fever is supported by a morbid irritability of the arterial system, or by a remote irritation originating from organic affections, tuberculous deposits in the lungs, etc.

Inflammations are successfully treated with Digitalin, in which affections it proves highly beneficial, both on account of its peculiar sedative influence over the arterial system, and its power of stimulating the absorbent vessels to action. In hypersthenic inflammations, arising from an exalted condition of the blood, other remedies will of course be needed to remove the cause of the disease, such as Podophyllin, Asclepin, and Veratrin, after which the Digitalin may be used as a palliative to quiet the irritable condition of the arterial vessels. But in vegetative inflammations, and such as are disposed to terminate in exudations or effusions, particularly when located in the serous membranes, as the pleura and peritoneum, or in the glandular structure, as the lungs, liver, etc., the Digitalin may be employed alone.

Digitalin is sometimes employed in acute dropsies of the cavities of the brain, but should never be given in sufficient doses to produce its narcotic effect. If used at all, small doses only should be employed.

In croup, Digitalin acts too slowly to be a certain and effective remedy, but is useful in the convalescing stages to prevent a relapse.

Digitalin is of excellent service in the treatment of puerperal fever, when the exudative inflammation of the peritoneum is distinctly manifest. In this affection the tincture may be applied locally with advantage, in connection with the internal use of the Digitalin.

Phlegmasia dolens and erysipelas are successfully treated with Digitalin in connection with Podophyllin. Digitalin may also be employed in some forms of hemorrhage, particularly those cases which are supposed to arise from a morbid irritability of the arterial system or some of its branches, and when organic diseases of the heart, lungs, or other organs exist, whereby the freedom of the circulation is interrupted. In hemoptysis and incipient phthisis pulmonalis, and for the suppression of colliquitive hemorrhoidal discharges, the Digitalin has been employed with much benefit. As a remedy for threatened abortion, arising from sanguineous congestion of the uterus, Digitalin, combined with Hyoscyamin and alternated with stimulants, such as camphor, etc., has been found of great service.

In organic and other abnormal affections of the heart and larger arteries, we have, in Digitalin, even in the most severe and malignant cases, an excellent palliative remedy. But in hyper-inflammation of these organs Digitalin may prove hurtful instead of beneficial, unless its employment be preceded by the judicious administration of Podophyllin. Digitalin relieves the asthmatic and syncoptic symptoms which are always connected with organic disease of the heart, and removes the chronic inflammation existing in the diseased parts, particularly of the serous membranes with which the interior of the heart and larger arteries is lined. It likewise promotes absorption and so lessens the tendency to exudation and effusion, particularly those dropsical effusions which so frequently occur as the sequents of organic disease. In these cases the Digitalin should be given in small and repeated doses.

In dilatation and aneurism of the heart, the Digitalin requires to be given in larger doses and alternated with tonics, as Iron, etc. In carditis polyposia, palpitation caused by morbid irritability, and pulsations felt in the abdomen, Digitalin is employed with much success. Also for the relief of angina pectoris or sternocardia.

Digitalin is extensively employed in the treatment of dropsical affections. This remedy is particularly indicated in those cases where exhalation is in excess of absorption, produced by erethism of the arterial system or of its extreme exhaling branches; as, for instance, acute dropsies following acute exanthema, as measles, scarlatina, which are mostly of an erethismal character, and the acute dropsies produced by sudden colds, particularly anasarca.

Digitalin is also of great value in the treatment of chronic dropsies, such as originate from a torpid or inactive condition of the absorbent and lymphatic systems and veins; as, for instance, chronic hydrocephalus, chronic hydrothorax, chronic ascites, etc. When there is great exhaustion and vital debility, Digitalin is contra-indicated. If employed at all, it must be in connection with stimulants and tonics.

In the asthenic form of dropsies common to aged persons, Digitalin may be combined with Hydrastin, Cerasein, etc., in conjunction with which it will be found serviceable in hydrothorax.

No other remedy has been more frequently employed in the treatment of phthisis pulmonalis than the Digitalin, yet it is by no means a specific. It is of great value as a palliative in tuberculous disease of the lungs, as it abates vascular excitement, stimulates absorption, and lessens the secretions of the bronchial mucous membranes. It is supposed to be capable, in some cases, of preventing tuberculous deposits.

Digitalin is of service in controlling the pneumonic symptoms accompanying phthisis pulmonalis. It arrests hemorrhage, abates the febrile symptoms, and removes the pulmonary and pectoral congestions. In these cases it should be given in small doses two or three times per day, occasionally omitting its use for a few days, and then resuming again. It may be combined, as circumstances require, with Hyoscyamin, tonics, and alteratives.

Digitalin is also successfully employed in the treatment of chronic pneumonitis and catarrhal complications, characterised by a continued sthenic irritability of the mucous membranes, and a tendency to exudations and effusions. Also in those chronic rheumatic affections of the lungs and pleura which so frequently terminate in hydrothorax. In these affections it is advantageously joined with Asclepin, Veratrin, Podophyllin, Hyoscyamin, etc. In phthisis laryngea and trachealis, arising from a strumous diathesis, the Digitalin may be given in small, repeated doses, combined with Asclepin, Prunin, or Rhusin, and alternately with Podophyllin, Phytolacin, etc.

We have in Digitalin an excellent remedy for scrofulous affections, particularly in persons of a full, plethoric habit, wherein excess of nutrition and repletion argue a torpid or inactive condition of the lymphatic system. It is also useful in the treatment of chronic scrofulous inflammations of the mucous membranes, strumous opthalmia, and in lingering scrofulous inflammations of the mesenteric glands. Digitalin has also been recommended in bronchocele.

The employment of Digitalin in nervous diseases cannot be recommended upon rational principles. It is sometimes employed in convulsive affections of the pectoral organs, as sternocardia, asthma, hooping cough, etc., and in convulsions, epilepsy, mania, hypochondria, paralysis, vertigo, amaurosis, etc.; but so long as we have better and safer nervines, the employment of the Digitalin should be limited so long as other complications do not positively indicate its use.

Digitalin is contra-indicated in violent and excessive sanguineous inflammations, vascular repletion, orgasm of the blood, extreme sensibility of the nervous system, great debility of the digestive apparatus, and true vital debility or atrophy.

Externally, Digitalin is employed in the treatment of scrofulous ulcers and tumors, local effusions of water, scrofulous diseases of the bones and joints, chronic exanthemas, psoriasis, etc. It may be dissolved in alcohol or made into an ointment with lard.

The average dose of the Digitalin is ONE FIFTH of ONE grain. In some cases it may be profitably increased to ONE HALF of ONE grain. But we profess only to approximate the quantity requisite in ordinary cases. We would advise the practitioner to always commence with *small doses*, and after a suitable time to increase, if occasion requires. Great caution should be exercised in its administration, and its exhibition never entrusted to unskillful hands. Above all things be sure to neutralise undue acidity of the stomach previous to its administration, and to render it as diffusible as possible by the free use of diluents. By so doing the danger of **cumulative action may** be avoided.

## CONCENTRATED TINCTURE DIGITALIS PURPUREA.

Properties and employment same as above. The strength of the Con. Tinc., as compared with the Digitalin, is as EIGHT to ONE; that is, EIGHT drops of the tincture represent ONE grain of The Digitalin. The dose will therefore vary from ONE to FOUR drops, in order to bear a relative proportion to the Digitalin. The tincture may always be relied upon as of definite strength, as it is prepared strictly in accordance with the principles recorded in the first part of this volume.

The tincture is convenient for external application, for which purpose it should be diluted with from four to eight parts of alcohol. It is of service as a topical remedy in local œdema, tumors, enlarged glands, etc. In the treatment of ovarian dropsy and ascites, we employ it in combination with tincture of Squills, as follows:

℞.
 Con. Tinc. Digitalis ................... ℨij.
 Tinc. Squills,
 Alcohol .......................... aa. ℨIV.

Mix. Bathe the parts freely three times per day, or apply cloths wetted with the liniment.

For internal use, when indicated, it may be combined with the Con. Tinc. Veratrum Viride. When astringents are indicated, with Con. Tinc. Rhus Glabra. When stimulants are needed, with Con. Tinc. Xanthoxylum Frax. As a general thing, however, it will be best to alternate the Tinc. Digitalis with tonics, stimulants, and alteratives, when such auxiliaries are indicated. When Asclepin is indicated, it should be reduced to solution, and the Tinc. Digitalis added to each dose as occasion requires. The conditions requiring the employment of either of the above named adjunctives have been pointed out in the preceding pages. The history there detailed of the properties and employment of the Digitalin is a faithful record of personal experience in its employment through a series of years, wherein both its advantages and disadvantages are fully explained.

# RHUSIN.

Derived from *Rhus Glabrum.*
Nat. Ord.—*Anacardiaceæ.*
Sex. Syst.—*Pentandria Trigynia.*
Common Names.—*Sumach, Upland Sumach, etc.*
Part used.—*Bark of the Root.*
No. of Principles, *two,* viz., *resinoid and neutral.*
Properties—*Tonic, astringent and antiseptic.*
Employment.—*Diarrhea, dysentery, apthous and mercurial sore mouth, diabetes, leucorrhea, gonorrhea, hectic fever, and scrofula.*

RHUSIN may justly be classed amongst the most valuable of the astringent tonics. It exercises a peculiar sanative influence over mucous membranes, and is invaluable in the treatment of many forms of disease affecting those surfaces. Being powerfully anti-septic, it is particularly useful in all cases manifesting a tendency to putrescency.

In diarrhea and dysentery, after the morbid accumulations have been removed by appropriate remedies, and the sthenic symptoms are measurably controlled, the Rhusin will be found of essential service in restraining and toning the action of the bowels. For this purpose it may be given in doses of

## CONCENTRATED MEDICINES PROPER. 215

TWO grains, once in two hours. When desired, it may be joined with other astringents, as the Geranin, Myricin, Lycopin, etc.; or with diaphoretics, as the Asclepin; or with stimulants, as the Xanthoxylin; or with tonics, as the Cornin, Cerasein, Fraserin, Eupatorin Perfo.; or with laxatives, as the Leptandrin, Euonymin, Juglandin; or with alteratives, as the Alnuin, Corydalin, Irisin, Stillingin, Phytolacin, Menispermin, Chimaphilin, etc. By judiciously selecting the adjunctive, combinations may be effected suited to the cure of the various diseases mentioned **at the head of** this article. Thus in diarrhea and dysentery, we combine it with Geranin, as follows:

℞.
   Rhusin,
   Geranin .............................aa. ℈j.

Mix, and divide into twenty powders. Give one every one to three hours, according to the urgency of the symptoms. If there is still a slight febrile condition remaining, we join it with Asclepin:

℞.
   Rhusin,
   Asclepin......................aa. grs. XV.

Mix, and divide into ten powders. Dose—same as above. For hemorrhage of the lungs, stomach, or bowels, we combine it with Lycopin:

℞.
   Rhusin,
   Lycopin .............................aa. ℈j.

Mix, and divide into twenty powders. Dose—one, every twenty or thirty minutes, until the hemorrhage is restrained, then at intervals of from one to three hours, and continued until the symptoms are fully abated. The same formula will be found of exceeding utility in the cure of diabetes. In this complaint the remedy may be administered three times per day. The dose will also require to be increased, in some cases, to double the quantity. When the system has been exhausted by profuse colliquitive discharges, and a relaxed condition of the bowels remains, Fraserin will be the best adjunctive:

℞.
    Rhusin........................grs. XV.
    Fraserin.......................grs. XXX.

Mix, and divide into fifteen powders. Give one every four or six hours, as occasion requires. In the treatment of leucorrhea, if constipation be present, the Rhusin may be given in TWO grain doses three times per day, and from TWO to FOUR grains of Leptandrin at bed time. Or they may be combined and formed into pills, as follows, although we prefer alternation :

℞.
    Rhusin........................℈j.
    Leptandrin....................℈ij.

Form a mass with mucilage of gum arabic, and divide into twenty pills. Dose—one, three times per day. Should they not prove sufficiently laxative, a dose of Podophyllin should be occasionally given at bed time. For gonorrhea, combinations may be effected with other of the vegetable alteratives, which, as we shall have occasion to so frequently mention them, it will not now be necessary to speak. We will say, however, that the Rhusin will be found a remedy of great utility in that complaint.

But the remedial value of the Rhusin is best displayed in the treatment of apthous and mecurial affections of the mucous surfaces. The various forms of stomatitis afford a wide range for its employment. It should be given in doses of TWO grains every four or six hours, and the mouth and fauces frequently gargled with a solution of the same. For the latter purpose, ONE DRACHM may be added to HALF A PINT of boiling water. We know of no more useful agent in the treatment of the distressing sequela that sometimes follow the use of mercurials. In case the lower portion of the alimentary canal be involved, the Rhusin may be administered by enema with advantage.

℞.
    Rhusin........................ʒj.
    Boiling Water.................Oj.

Of this infusion, from TWO to FOUR ounces may be administered, tepid, every two to four hours. The same will be found exceedingly efficacious in some cases of dysentery and rectal hemorrhage. Some practitioners, in the above complaints, combine the Rhusin and Myricin in equal proportions.

Rhusin has been employed with advantage in hectic fever, in which complaint it may be sometimes beneficially joined with Digitalin, as mentioned under that head. In scrofula also, particularly those cases involving the mucous surfaces, the Rhusin has been found valuable. In such cases it should be alternated with alteratives and tonics. In the diarrhea of typhoid fever, and in all cases where a putrescent tendency is manifest, the Rhusin will be found a reliable and appropriate remedy. When astringent, tonic, anti-septic, and stimulant properties are indicated, a combination of Rhusin with Xanthoxylin will be found equal if not superior to any other. The latter two remedies act admirably together, and indications for their employment will be met with in diarrhea, dysentery, cholera infantum, typhoid fever, scarlatina maligna, etc. In ulcerations of the stomach and bowels, the Rhusin should not be omitted. Average dose, TWO grains.

## CONCENTRATED TINCTURE RHUS GLABRUM.

Properties and uses same as the preceding. Average dose, THREE drops. Convenient for combining with other of the concentrated tinctures, when auxiliary properties are indicated. For example:

℞.
    Con. Tinc. Rhus Glab.
    Con. Tinc. Digitalis Purpu.

℞.
    Con. Tinc. Rhus Glab.
    Con. Tinc. Senecio Gracil.

℞.
    Con. Tinc. Rhus Glab.
    Con. Tinc. Xanthoxylum Frax.

℞.
    Con. Tinc. Rhus Glab.
    Con. Tinc. Smilax Sarsa.

The average doses being given under the proper heads, the proportions may be easily regulated.

# BAPTISIN.

Derived from *Baptisia Tinctoria.*
Nat. Ord.—*Fabaceæ.*
Sex. Syst.—*Decandria Monogynia.*
Common Names.— *Wild Indigo, Horsefly Weed,* etc.
No. of Principles, *two,* viz., *resin* and *neutral.*
Properties.—*Alterative, emetic, laxative, stimulant, emmenagogue, tonic,* and *antiseptic.*
Employment.—*Amenorrhea and defective menstruation, erysipelas, hepatic disorders,* whenever an *alterative* is indicated, and in *scarlatina and typhoid fevers,* and in all diseases that have *a putrescent tendency.*


BAPTISIN is possessed of more energetic emmenagogue properties than the plant has generally been accredited with. We have employed it with gratifying success in the treatment of amenorrhea and defective menstruation. Also in cases of vicarious menstruation, in combination with Podophyllin, with signal success. The average dose of the Baptisin is two grains. The dose may be repeated twice or thrice a day as circumstances require. In too large doses it will produce nausea, emesis, and catharsis. In the treatment of amenorrhea and defective menstruation, the Baptisin should be given in doses of from ONE to THREE grains three times per day, and a dose of Podophyllin and Leptandrin administered once or twice a week at bed time. In the treatment of vicarious menstruation, particularly those cases accompanied with periodical diarrhea, we have found the following combination entirely successful, when administered during the intermenstrual period:

℞.
    Baptisin ............................ ℈j.
    Podophyllin........................grs. X.
    Caulophyllin ......................grs. XV.

Mix, and divide into ten powders. Exhibit one every night, or every other night, according to the condition of the bowels. The quantity of Podophyllin should be sufficient to produce a mild cathartic effect at first, and afterwards the quantity may be reduced so as just to secure its alterative and laxative effect. When necessary, it may be alternated with tonics, as Helonin and Iron, or Cerasein.

In erysipelas the alterative and antiseptic properties of the Baptisin make it a remedy of great value. It may be administered, internally, in doses of from ONE to TWO grains once in four hours, and if there be ulcerations or sloughings, the parts should be covered with dry Baptisin, over which, if there be much pain, heat or inflammation, place the cold water bandage. This we have frequently employed and found effectual. The application of the Baptisin may be repeated two or three times a day, and the bandage re-wetted as often as it becomes dry or much heated. The same treatment will be found of essential service in other forms of acute as well as of chronic exanthema.

Baptisin is a sure and powerful alterative, and may be employed with confidence in all affections of the glandular system. In hepatic derangements it will be found a valuable auxiliary, and in a great many cases may be depended upon alone. In scrofula and cutaneous disorders, few remedies are more beneficial. In these cases it should be given in small doses, and its use persisted in for a length of time. It should be alternated with occasional doses of Podophyllin and Leptandrin, and also with other alteratives. In consequence of the stimulant properties of the Baptisin, it is valuable in all cold and indolent conditions of the system, such as usually accompany scrofula, white swelling, hip disease, scaly eruptions of the skin, etc. Many valuable combinations may be effected with other of the concentrated agents, as the circumstances of the case may indicate. In the treatment of ulcerative inflammations of the stomach and bowels, and chronic diarrhea, and dysentery, its use should never be omitted. We consider its tonic and antiseptic properties as of paramount value, and as specially indicated in all cases of internal ulcerative inflammations, putrescency, gangrene, etc. In the various forms of stomatitis, mercurial sore mouth, putrid sore throat, scarlatina maligna, typhoid fever, dysentery, and inflammation of the bowels, we have, in the Baptisin, one of the most powerful, and, at the same time, safest antiseptic remedies in the range of the Materia Medica. If astringent properties are indicated in connection with the Baptisin, we have Geranin, Myricin, Rhusin, Lycopin, Trilliin, etc. If diaphoretics are needed, Asclepin, Veratrin, etc. If more stimulating properties are required, Xanthoxylin. Of alteratives we have, as adjunctives, Alnuin, Chimaphilin, Rumin, Irisin, Phytolacin, Stillingin, Smilacin, etc. To increase its laxative property, Euonymin, Hydrastin, Menispermin, Apocynin, Leptandrin, Podophyllin, etc. Combined with Caulophyllin, we have found it very serviceable in certain forms of dyspepsia, particularly those cases accompanied with irritability of the stomach, acid eructations, griping pains and looseness of the bowels, with frequent, small and offensive stools. In a majority of cases it is better to precede the administration of the Baptisin and Caulophyllin with a cathartic dose of Podophyllin and Leptandrin, in which the latter should largely predominate.

With Leptandrin the Baptisin will be found excellent in chronic affections of the liver, accompanied with constipation. We combine as follows:

℞.
    Baptisin ................................ ℈j.
    Leptandrin ............................. ℈ij.

Form a mass with mucilage of gum arabic, and divide into twenty pills. Give from one to two, morning and evening. The same will be found excellent in chronic diarrhea, and dysentery, and ulcerations of the bowels. If a milder chologogue and laxative is required, substitute the Juglandin for the Leptandrin.

When astringents are indicated, we prefer the Rhusin, particularly in typhoid fever, mercurial ulcerations, etc.

℞.
    Baptisin ........................... grs. X.
    Rhusin ............................ grs. XX.

Mix, and divide into twenty powders. Give one, every one or two hours, according to the urgency of the symptoms.

In the treatment of virulent leucorrhea, the Baptisin will be found one of the most effective agents, to be used both internally and locally. For internal use we generally combine it with other remedies, as the Hydrastin, Helonin, Phytolacin, etc. Locally, the following:

℞.
    Baptisin ............................... ʒij.
    Boiling Water ......................... Oj.

Infuse the Baptisin in the water, and inject with a proper syringe three or four times a day. It may be used tepid or cold, as preferred. We frequently vary the prescription by combining the Baptisin with other agents; as follows:

℞
    Baptisin,
    Hydrastin ....................... aa. ʒj.
    Boiling Water ....................... Oj.

Or,

℞
    Baptisin,
    Myricin ......................... aa. ʒj.
    Boiling Water ....................... Oj.

Or,
℞.
    Baptisin............................... ʒj.
    Pul. Gum Myrrh ...................... ʒij.
    Boiling Water ........................ Oj.

Infuse and strain. The latter is excellent in ulcerations of the vagina, os uteri, and congestions and inflammations of the uterus and vagina generally, and for the relief of the irritation produced by acrid menstrual discharges.

In the treatment of apthous sore mouth and similar ulcerative affections, the Baptisin should be used as a gargle, of the strength of from ONE to TWO drachms to the pint of boiling water. In these cases it is better joined with Rhusin, ONE drachm of each to the pint. In severe cases double the quantity may be employed.

In combination with Dioscorein, the Baptisin will be found of great service in the treatment of a variety of intestinal affections, such as are accompanied with spasmodic pains, flatulence, and acrid fœcal discharges. It has also been found beneficial in pneumonia and chronic rheumatism. It excites the secretions of the glandular system generally, and of the liver and uterus particularly. In over doses it produces considerable prostration of the whole system, from which, however, the patient quickly recovers when the remedy is omitted. It should not be used during the period of utero-gestation, as it is capable of producing abortion, for which purpose we have known it to be used by quacks and empirics. The danger to the general health is very great when used in sufficient quantities to produce this result.

Externally, the Baptisin admits of a wide and beneficial range of application. Its peculiar antiseptic property renders it a valuable local remedy for erysipelatous and other ulcers, strumous and syphilitic opthalmia, otorrhea, ulcerated sore mouth and throat, chancres, ulcerations of the cervix uteri, sore nipples, mammary and other abscesses, inflamed tumors, and in all affections having a gangrenous tendency. To open ulcers, the dry powder may be applied, as in erysipelas, scrofulous ulcers, ulceration of the cervix uteri, etc. For ophthalmia, otorrhea, etc., it made be made into decoction, from ONE to FOUR drachms being added to a pint of boiling water. The same will answer for injections into mammary and other abscesses, and for the relief of fetid vaginal discharges. As a local application to tumors and inflamed glands, it may be applied by means of a suitable poultice, as of elm or flax-seed, the surface of which may be sprinkled over with the Baptisin. In the same manner it may be applied to open ulcers. For the treatment of scaly eruptions of the skin, it may be dissolved in alcohol, ONE drachm to FOUR ounces, or made into an ointment with lard, ONE drachm to the ounce. As a safe and reliable antiseptic, **it is worthy the entire confidence of the profession.**

# PODOPHYLLIN.

Derived from *Podophyllum Peltatum.*
Nat. Ord.—*Berberidaceæ.*
Sex. Syst.—*Polyandria Monogynia.*
Common Names.—*Mandrake, May Apple,* Wild *Lemon*
Part Used—*The Root.*
No. of Principles, *three,* viz., *resinoid, alkaloid* and *neutral.*

Properties—*Emetic, cathartic, chologogue, resolvent, alterative, diuretic, diaphoretic, emmenagogue, vermifuge, revellent, etc.*

Employment—*Fevers* and *inflammations* of almost every type, all disorders of the *liver, spleen,* and other viscera, *croup, pneumonia, rheumatism,* both acute and chronic, *scrofula, indigestion, venerial diseases, jaundice, piles, constipation, dropsy, gravel, inflammation* of the *bladder, suppression* and *retention* of the *urine, eruptions of the skin, amenorrhea, leucorrhea, opthalmia, otorrhea,* and, in short, whenever **an** *alterative* is required

In essaying to treat upon the properties and employment of this truly invaluable remedy, our mind misgives us upon two points; first, as to whether we shall be able to adequately express our knowledge and convictions of its utility; and, secondly, if enabled to do so, whether our statements will receive that credence to which they are entitled, or be passed over with that indifference which too frequently characterizes minds immured in their own self-sufficiency. Nevertheless, we shall endeavor to fully, fairly, and truthfully detail such positive knowledge as may be in our possession, drawn from the private resources of personal clinical experience, and from the public acknowledgements of writers held in high estimation by the profession, relying upon the capability of the remedy to accomplish all that we shall claim for it. Were mankind as ready and willing to investigate, comprehend, appreciate, and acknowledge, as they are to doubt, disbelieve, condemn, and repudiate, there would be more truth and harmony in the affairs of life. Education, habit, custom, begetting as they do a reprehensible confidence in, and slothful dependence upon the sayings, doings, doctrines, and practices of former ages, form a sad bar to the progress of innocuous medication. We are among those who believe that a benign and all-wise Creator has endowed the earth with inexhaustible resources of means wherewith to meet all the necessities of its children; and those of a kind ever conservative to the integrity and duration of the objects upon which they are employed. It is in this light that we look upon the Podophyllin and kindred remedies, holding the sentiment that all remedial agents should be always conservative, and never destructive in their influences. A better knowledge of such means is being opened up by the progressive enlightenment of the human mind, and the profession are beginning to understand and appreciate the nearer compatibility of organic medicines with the functions of organic life.

The Podophyllum Peltatum has been long and favorably known, in the crude state, as an efficient remedy in disorders of the liver. Much error, however, pertains to many written histories of the plant. Many writers have likened its properties to those of Jalap, deeming the two nearly or quite analogous. No greater misconception could possibly be made in relation to the remedial properties of the plant. Jalap is simply an irritating hydrogogue cathartic. Its history is told in a single line. Not so with the article under consideration, as we shall have occasion to show.

In relation to the character and number of the proximate principles upon which the plant depends for therapeutic value, much ignorance has prevailed, and still prevails, even among many manufacturers engaged in preparing concentrated organic remedies for the use of the profession. One offers us an *alkaloid* Podophyllin, another a *resinoid* Podophyllin, and so on; but none give us a true account of the chemical constituents of the plant, in fact give us no explanation at all, except that they have obtained a precipitate which they have dried down to a powder, and which they *guess* is *the* active principle of the Podophyllum, and as such they represent it to the profession. We have before explained, and deem it not out of place to reiterate, that the therapeutic properties of the Podophyllum Peltatum reside, not in *one*, but in *three* distinct and separate proximate principles, each one of which represents its individual share of the aggregate remedial virtues of the plant. These three principles are termed resinoid, alkaloid, and neutral. The resinoid represents the emetic, cathartic, and chologogue properties chiefly. It is composed, as heretofore stated, of a number of distinct resins, each one possessing a different degree of electro-negative reaction. We have separated the resinoid of the Podophyllum into five different resins, and have reason to believe that a still greater complexity exists. It possesses a degree of escharotic power, and when applied externally to fungous growths, will dissolve them down. It produces, however, too much inflammation to render it a desirable escharotic. Combined with sulphate of zinc and Hydrastin, it has been found valuable as an application to cancerous growths. As a counter-irritant, dissolved in alcohol, is is one of the most active and efficient that we have ever employed. It produces a rapid pustulation, which appears first in the form of minute vesicles filled with a serous fluid, which speedily changes to a whitish or yellowish pus. The superficial inflammation is at the same time quite severe. The pustules, as a general thing, are slow in healing. We employ it in chronic and obstinate cases of local neuralgic pains, spinal irritation, chronic hepatitis, pleuritis and synovitis, morbus coxarius, etc.

In the alkaloid and neutral principles we have the diuretic, diaphoretic, alterative and laxative properties of the plant in an eminent degree. They also possess a considerable degree of chologogue power, and seldom prove emetic. These two last mentioned principles exercise a wonderful modifying power over the action of the resinoid principle. None but those who have tested the matter can appreciate the great difference between the physiological impressions of the resinoid when used alone, and those of the three principles combined. Many who have deemed the resinoid Podophyllin too harsh and drastic, and justly so, have found the combined principles to answer all their expectations. We earnestly invite the attention of the profession to the explanations we have given in reference to the multiplicity of principles residing in the Podophyllum Peltatum, and, if doubtful of the correctness of our statements, to put us to whatever test may be deemed necessary. We have no mercenary motive to subserve in our essay upon this article, neither in aught we ever have or ever shall submit to the profession, hence fear not for the results of the severest criticism. We desire investigation and scrutiny, in order that the profession may become enlightened against the errors and frauds of ignorant and incompetent manufacturers of concentrated organic remedies; and in order that the justice of our claims to a truthful exposition of the number, character and properties of the proximate active principles of plants may be vindicated.

We have, in the Podophyllin under consideration, a complete and reliable substitute for mercury and its preparations. The plea that the vegetable kingdom affords no remedy of equal fficacy with calomel and other mercurials in disorders of the liver, and in all cases in which those preparations are employed, is no longer tenable. Podophyllin has been called the "Vegetable Calomel." So far as the similitude relates to its power to produce sanative results, it is correct; but here the resemblance ceases. For all the *good* that calomel can possibly do, the Podophyllin is equally competent, while at the same time its operation is entirely devoid of those unfortunate results which so often follow in the wake of its mineral protonymic. It may seem presumptuous in us to advocate an equality between a remedy of comparatively recent discovery, and one which has received the sanction of the profession for nearly four hundred years, yet if we can succeed in showing that the Podophyllin will effectually subserve all the curative purposes of mercury, and is, at the same time, innoxious in itself, we trust that we shall not be deemed hasty or incautious in our advocacy of a substitute.

It has been said that Podophyllin is capable of producing ptyalism, but we have never seen any evidence of the fact in persons who had *never taken mercury*. The only symptoms of salivation we have ever observed have been in those cases where mercury had been taken at some previous time. Podophyllin is powerfully resolvent, and by its peculiar excitation of the glandular system will sometimes dislodge deposits of latent mercurial atoms, and so bring about a season of mercurialisation. Lobelia, Irisin, Phytolacin, etc., will frequently do the same. We believe it is conceded by the most intelligent writers and teachers of the present day, that the production of ptyalism is entirely unnecessary to the cure of disease, hence the absence of this power in the Podophyllin does not militate against its value. We have frequently induced a degree of salivation in patients by passing a current of electro-galvanism through the salivary and cervical glands, but only in those cases where mercurials had been previously administered. The effect in these cases was produced by the dislodgment of mercurial deposits, and as soon as they were removed the glandular inflammation would subside, nor would the re-application of the electricity ever again induce a similar train of symptoms.

In large doses, say from THREE to FIVE grains, Podophyllin is an active emeto-cathartic. Its operation is attended with copious bilious discharges, a lingering, death-like nausea, and frequently with severe griping pains in the small intestines. The primary impressions of Podophyllin are expended upon the gastro-enteric and hepatic apparatus, and nausea and vomiting seldom occur until from two to four hours after the medicine has been administered. From this fact it may be learned that the sickness, griping, and other unpleasant symptoms arise more from the acrid character of the morbid matters dislodged, than from the primary influences of the remedy itself. The neutral and alkaloid principles are completely soluble in the stomach, while the resinoid principle is soluble only in the enteric secretions. For a fuller explanation of the action of the different principles, the reader is respectfully referred to page 85, *et seq.* If Podophyllin be retained for three quarters of an hour after it is administered, it will not be rejected by vomiting, showing that within this period it has entered into solution and passed into the circulation, which fact will be manifested by its producing its characteristic influences upon the system, even though free emesis occur immediately upon the termination of this period. The therapeutic action of Podophyllin is completely suppressed by the presence of a considerable quantity of *lactic acid*, but operates without hindrance in the presence of *acetic acid*. Hence the necessity of neutralising undue acidity of the stomach previous to its exhibition will be apparent, as well as to avoid the use of such substances as will give rise, by putrefactive decomposition, to the formation of lactic acid. Sugar is particularly objectionable in connection with Podophyllin. We have previously shown that sugar, when in solution and exposed to a temperature above 80° of Farenheit, undergoes a putrefactive fermentation, and gives rise to the formation of a number of products, among which is lactic acid. Hence the use of syrups, sweetened infusions, etc., should be dispensed with while the system is under the influence of Podophyllin. Were the sugar properly digested, it would be of no disadvantage;

but those conditions requiring the exhibition of Podophyllin are unfavorable to the digestion of nutritious matters of any kind, and much more so when the digestive apparatus is under its immediate influence. The sanative impressions of Podophyllin upon the digestive organs, unlike those of many other remedies, are indirect and subsequent to its specific constitutional influences. Digestion cannot proceed during the immediate operation of Podophyllin, nor until several hours have elapsed after its cathartic powers are manifested, when given in cathartic doses. Populin, Xanthoxylin, and other stimulants and tonics, on the contrary, directly promote digestion, hence are given with the greatest advantage immediately before or after meals, in order that their specific influence may be expended upon the digestive organs at the precise time when extraneous aid is necessary.

Chloride of sodium, common salt, enhances the activity of Podophyllin, and to the abundant use of this condiment may be attributed the apparent hyper-cathartic effect sometimes observable in the use of this remedy. Our attention was first called to this fact some five years since, and the phenomenon was at first ascribed to the eating of oysters, but subsequent observations demonstrated the fact that it was the salt so conveyed into the system that produced the effect. This property of salt renders it valuable in promoting the action of Podophyllin in those cases where great coldness and torpidity exist, and when that remedy is tardy in operating. In all cases of a sthenic character, however, salt should be used in moderation while the system is under the influence of Podophyllin. We generally confine our patients to a diet of simple corn meal gruel for a period of twenty-four hours after exhibiting a full dose of Podophyllin. If it be desirable to promote the action of the medicine, salt may be added to the gruel in sufficient quantity to produce the desired effect.

Many suggestions have been made in regard to the combination of other agents with the Podophyllin, in order to modify its operation. Among those agents, we may mention Leptandrin, Jalapin, Asclepin, Caulophyllin, Gelsemin, Phytolacin, etc. The Leptandrin is, perhaps, more employed than any other. There is no doubt but that it both enhances and modifies the chologogue power of the Podophyllin, while at the same time it lessens the intestinal irritation. It also seems to be of great service in securing the full alterative influence of the Podophyllin, although a portion of this influence is undoubtedly due to the adjunctive itself. In typhoid fever, dysentery, and other diseases attended with intestinal irritation, we deem the Leptandrin an indispensible auxiliary. We usually employ two parts of Leptandrin to one of Podophyllin,

Jalapin with Podophyllin is indicated in dropsy, and in all cases where a speedy evacuation of the immediate contents of the bowels is desirable. The Jalapin will neither quicken nor in any other way influence the action of the Podophyllin, which will manifest its accustomed influences independent of the Jalapin. In congestions of the portal circle, accompanied with intestinal engorgement, the combination of Jalapin with Podophyllin is appropriate. By the use of the Jalapin in these cases, we get a prompt evacuation of the alimentary canal as the result of its more speedy local cathartic power. But the Podophyllin will take its own time, and its general influence will be the same as if no Jalapin had been employed. In the treatment of dropsies, we have derived more prompt and permanent sanative results from a combination of Podophyllin, Jalapin, and Cream of Tartar, than from any other hydrogogue remedy.

Asclepin has long been a favorite adjunctive to the Podophyllin, with us. Long before the discovery of the active principles of these plants, we were in the habit of combining the crude Asclepias with the Podophyllum. It lessens the tendency to griping, and by virtue of its diaphoretic properties, seems to enhance the influence of the Podophyllin upon the sub-cutaneous glandular structure. For this reason we deem it a valuable adjunctive to the Podophyllin in the treatment of cutaneous diseases. Also in all affections attended with febrile symptoms wherein Podophyllin is indicated.

Caulophyllin is also an excellent modifying agent for combining with the Podophyllin. Its anti-spasmodic properties are useful in controlling the tendency to nausea, pain and spasm. It is particularly serviceable as an auxiliary in the treatment of amenorrhea, hysteria, chorea, and all nervous affections. Also in certain forms of indigestion, cholera morbus, etc.

Gelsemin is used for the same purposes as the above. It is a more enegetic anti-spasmodic and relaxant, and at the same time possesses other properties frequently indicated in connection with the Podophyllin. We are in the habit of prescribing in combination with the Podophyllin daily. In hepatic congestions, the forming stages of fevers, pneumonia, croup, and whenever febrile and spasmodic symptoms are present, we seldom omit it. By relaxing spasm, abating febrile excitement, and soothing the irritability of the nervous system, it quickens and promotes the operation of the Podophyllin. We find it of great service as an adjunctive in a great variety of chronic diseases.

Phytolacin is peculiarly serviceable as an adjunctive in the treatment of obstinate hepatic disorders, constipation, and in all cases accompanied with a languid or torpid condition of the system. Whenever it is found difficult to bring the system under the constitutional influence of Podophyllin, by reason of excessive sluggishness or other causes, vital debility excepted, the Phytolacin will be found to answer an admirable purpose. In syphilis, scaly eruptions of the skin, chronic hepatitis, scrofula, etc., the Phytolacin will always prove a valuable auxiliary.

From five to ten grains of super-carbonate of soda may be advantageously combined with each dose of Podophyllin in case acidity of the stomach be suspected. Capsicum is a good adjuvant to Podophyllin in cold and languid conditions of the system. Many other combinations may be effected, some of which we shall have occasion to notice, and others will readily suggest themselves to the practitioner.

Of the special employment of Podophyllin in the treatment

of disease, we would mention fevers generally as affording frequent and decided indications for the use of this remedy. In the treatment of fever and ague, we almost invariably precede the employment of other remedies by the free exhibition of the Podophyllin. By so doing, in this climate, we cut the disease short at once, and oftentimes have no occasion for further medication. We have known many cases of intermittent fever to yield to a single dose of Podophyllin, and we have no doubt that the credit of cure is frequently due to this agent, when it is attributed to other means. In those cases of fever and ague in which the bowels are a special point of congestion, manifested by a troublesome and painful diarrhea, the Podophyllin is sometimes inadmissable. If, however, the diarrhea depend upon a functional disturbance of the liver, it will be indispensible. It should always be combined, in such cases, with Leptandrin and Caulophyllin or Dioscorein. If, on the other hand, the diarrhea arises from a primary intestinal congestion, and be of a serous or mucous character, the Podophyllin should be dispensed with, and the chief reliance be placed upon Leptandrin, or Euphorbin, in combination with diaphoretics and anti-spasmodics. In cases of this type, it will be better, as a general thing, to administer the above remedies in divided doses. The following formula is excellent:

℞.
    Leptandrin............................ ℈j.
    Asclepin,
    Dioscorein ........................aa. grs. X

Mix, and divide into ten powders. Give one every two hours until the alvine discharges assume a healthy appearance. Astringents may then be employed, but we seldom find them necessary. The above formula may be varied at the option of the practitioner. In the treatment of chronic cases of this complaint, in adults, we generally premise our subsequent treatment with the following somewhat heroic prescription:

℞.
    Podophyllin,
    Euphorbin ......................aa. grs. ij.
    Leptandrin ........................... grs. iij.

Mix, and give at a dose. This will produce free emesis and catharsis, and thoroughly arouse the system. If the first dose does not sufficiently break up the hepatic obstructions and awaken the system from its torpor, we repeat the dose at the expiration of from twenty-four to forty-eight hours. Of course the quantity of the ingredients in the above formula must be regulated to the necessities of the case in hand. In all cases attended with gastric or enteric irritation, a free use of mucilages and demulcents is advisable. The above prescription we have found of eminent service in the forming stages of bilious, typhoid, and other fevers, pneumonia, erysipelas, acute rheumatism, etc. We vary the formula to meet the indications. If a considerable degree of febrile excitement be present, we usually substitute from ONE HALF to TWO grains of Gelsemin for the Euphorbin, increasing, if necessary, the proportion of Podophyllin, or Leptandrin, or both. Congestion of the brain has frequently yielded to the prompt administration of this remedy. We have cured Panama fever of eight months duration by means of Podophyllin and Gelsemin, followed by Hydrastin and Xanthoxylin. We wish it distinctly understood, that the treatment here detailed applies to the peculiarities of this climate. We are aware that the habits of individuals, food, water, climatic and other influences all tend to modify both the types of disease and action of medicines, and that it is necessary to modify the combination of agents in accordance with the circumstances of their employment. These peculiarities it is the duty of the resident physician to ascertain, and, having made himself thoroughly acquainted with the therapeutic properties of the agents he employs, to modify his treatment accordingly. We have exhibited the Concentrated Medicines in the States of North Carolina, Alabama, Florida, and upon the Mississippi River, both to the white and colored races, and we never had them fail of their accustomed effect. Scarlatina, acute rheumatism, nephritis, diarrhea, dysentery, and other diseases yielded as readily to the organic remedies as in our native clime. Yet our residence in those localities was too brief to enable us to speak authoritatively as regards the proper plan of treatment to be there pursued. In miasmatic districts, as in the valleys and river bottoms of the West, disease assumes a more periodic type, and, in complaints like rheumatism, cholera morbus, etc., unless anti-periodics be promptly administered during the remissions, relapse will speedily follow relapse. From this fact we may learn the importance of using proper means to maintain a favorable condition when it is once brought about. Upon this point we shall have more to say when treating of special anti-periodics.

In many instances it will be necessary to combine Podophyllin with active stimulants, at other times with sedatives, diaphoretics, antispasmodics, or simply with mucilages or demulcents. We cannot undertake to point out all the specific indications in which these various modifications will be necessary, neither do we deem it necessary, as the practitioner cannot fail to comprehend the combination suggested by the circumstances of the case.

In the treatment of typhoid fever, the Podophyllin is sometimes deemed too irritating in its operation. Such, no doubt, is the case in many instances. We have heretofore spoken of the escharotic property of the resinoid principle of the Podophyllin, and we again desire to draw attention to the fact. It is all the more important to keep this fact in view, when we consider that the Podophyllin of many manufacturers consists of the resinoid principle alone, and we have no doubt but that this circumstance will account for the drastic effect observed by some practitioners in the operation of Podophyllin, and which, by them, has been justly considered an objectionable feature. We have before stated that Podophyllin is contra-indicated in gastritis and enteritis, and whenever there is evident local inflammation of any portion of the glandular structure of the intestines. It must be borne in mind, that organic substances are possessed of chemical affinities equally with the inorganic, and that in diseased conditions of any portion of the animal economy, not only is there a functional aberration, but also is the chemical constituency of the apparatus and its secretions essentially changed and modified. With this change of constitution comes new affinities, and a substance which, under other circumstances, would pass harmlessly over a given surface, is, by the consequent reaction resulting from this changed composition of the secretions, converted into a drastic irritant. It might be said that these phenomena arise from a modification of nervous impressibility, but we invariably find that such modification is attended with change both of the chemical structure of the organ and its secretions. It is important, therefore, if Podophyllin be employed at all, that it be so combined that these accidents of impression be obviated. In the treatment of Typhoid Fever, and other acute diseases, when called in the advanced stages, if we find on examination a suppression of the mucous secretions, we do not immediately administer Podophyllin, be it, in other respects, ever so much indicated. Our first reliance is upon diluents and demulcents, preferring those of a diaphoretic character, in order that a degree of reparation may be made for the expended fluids of the system. As soon as we have awakened the secretive action of the mucous surfaces, we administer our Podophyllin, or whatever other constitutional remedy we may deem necessary in the case. It is bad practice, when the tongue, mouth and fauces are dry, parched and inflamed, showing, evidently, a suspension of action on the part of the exhalents, to administer Podophyllin or any other remedy requiring the menstrua of solution, and which are capable, if they remain undissolved and unabsorbed, or even if they dissolve very slowly, of expending an unneeded and undesirable local influence. Nothing is more essential to health than that a proper diluency of the blood and various juices of the system be maintained. The very suspension of the exhalations of the serous and mucous membranes is oftentimes a conservative manifestation on the part of the system, showing that the dissipation of the fluids has reached an extent inconsistent with the integrity and duration of the animal economy. It is in cases like these that the very blood corpuscles themselves become shrivelled and shrunken, having, by the action of exosmose, given off a portion of their water to supply the demand denied from the proper sources. Not many years have elapsed since the standard treatment of patients laboring under febrile forms of disease was such as to consume them by a slow process of moist incineration. Venesection, evacuants, and other artificial means of depletion were employed, while, at the same time, the patient was denied the indulgence of that indispensible and Heaven sent conservator, water, even while the body shrunk and consumed in the pyrexian furnace. Bleeding, blistering, cupping, leeching, vomiting, purging, sweating and diuresis, served to aid the fever in extracting and dissipating the fluids of the body, leaving the vital currents to thicken and stagnate in the channels of life, and planting the banner of death at the very citadel of the life forces. Fortunately for the interests of suffering humanity, a reform in this respect is apparent amongst the more intelligent of the profession although we fear that the requirements of the natural laws are not, in many instances, sufficiently regarded.

Water is, properly speaking, the only diluent. At the same time it is capable of holding in solution certain therapeutic principles which act as stimulants, both upon the exhaling and absorbing vessels. Hence, by the administration of infusions of some of the simpler plants, such as yield soluble neutral principles possessed of diaphoretic properties, we may at the same time furnish the material for maintaining a proper diluency of the various juices, and the means conducing to its appropriation. When attainable, we should scarcely make use of any other remedy than the Asclepin for that purpose, deeming it always appropriate. As it is nearly all soluble in warm water, it is of convenient and admirable utility. Mucilages and demulcents act in a manner mechanically, shielding the irritable and irritated membranes from the action of the acrid secretions, and at the same time apparently soothe and allay the excited condition of the mucous surfaces. They also, as a general thing, afford absolute nutriment, and, provided the system be in a condition to appropriate nutritive matters, will answer both as food and medicine. It is necessary to success-in the practice of medicine to always bear in mind the fact, that medicines calculated to induce constitutional changes are always first *acted upon* by the system, and that the different degrees of ability on the part of the system to properly dispose of or appropriate a remedy, will regulate in the same degree its positive influences; while the absence of this power will simply afford negative results. This is precisely the case with food, and, in chronic diseases, when the nutritive apparatus fails to make disposition of the aliment taken into the stomach, we need scarcely hope that medicines will share a better fate. How important, then, that due discrimination should be exercised in selecting the various remedies used in the cure of disease, always keeping in view the question of adaptation on the part of the remedy, both as regards its therapeutic and physical character, to the conditions present. It is in consequence of this constitutional diversity that individual remedies cure in some cases and fail in others, or exhibit various shades of curative power.

We hope we shall be pardoned for digressing somewha from the strict details of the remedy under consideration, but we could not well do otherwise than revert to a few general principles governing the successful employment of remedial agents, and especially the Podophyllin.

In the treatment of exanthematous fevers, Podophyllin is frequently indicated in the forming stages, and its prompt administration will deprive this class of diseases of much of their malignancy. If the symptoms indicate a considerable degree of hepatic derangement, it should never be omitted. Later than this, it is bad practice to administer Podophyllin, or indeed any other cathartic, until the efflorescence is complete and mature. At this stage, the Podophyllin will be found to act more desirably than any other agent of its class. In these cases it should, as a general thing, be combined with Leptandrin. In some cases stimulants may be indicated, as Xanthoxylin, Capsicum, etc. The general directions in the early part of this article may be consulted in regard to suitable combinations. In the treatment of fevers and other inflammatory diseases, a single dose of Podophyllin must not be relied upon in the outset, unless the alvine discharges give evidence of the removal of all morbid accumulations. Our practice in such cases is to repeat the Podophyllin once in twenty-four hours, or at such periods as may be suitable, until the discharges from the bowels give evidence of effective and complete depuration through that channel. Unless this be done, the practitioner will frequently fail of his objects. Evidence is thus afforded that the principal obstructions are broken up, and that the effete and corrupt materials which act as fuel to the flame are expelled. The success of the subsequent treatment will depend in a great measure upon the consummation of this result. Following this, diaphoretics, sedatives, febrifuges, nervines, etc., will act with greater promptitude and certainty, as they will not have to contend against the principal cause which perpetuates the functional disturbances, and frequently leads to organic lesions; but simply have to harmonise the disturbances remaining after the expulsion of a cause which has ceased to operate. Much less medicine will be needed subsequently, and greater certainty will attend its administration. Thus, in bilious, scarlet, and other fevers, if this be done, the subsequent employment of Veratin, Asclepin, Gelsemin, etc., will be attended with more speedy and satisfactory results; while, if this be neglected, and the direct cause of excitation be allowed to remain, seldom can a sufficient amount of calmative influence be brought to bear to harmonise the action of the disturbed functions. We hold it to be an axiom in medical science, that every *effect* in turn becomes a *cause*. Let us look a moment at the approach and progress of a case of typhoid fever. First we have slight debility or lassitude, a dull feeling in the head, followed by pain, aching, and lameness in the limbs, soreness of the flesh, appetite feeble or wanting, bowels generally constipated, skin dry, urine scanty, tongue slightly coated, taste impaired, accompanied with other symptoms and modifications which finally usher in a season of chills, alternated with febrile paroxysms until the disease is fairly established. Here we see that there has been manifest tardiness on the part of the depurating functions, the effect of which is *retention* of morbid and effete matters, which in turn results in *accumulations*. These retained and accumulated matters are acrid and morbific, as we may learn from the fact that nature frequently makes an effort to expel them in the earlier stages of the disease by diarrhea. But many practitioners thwart this early effort of the system by administering opiates and astringents. Our practice is different. We hold that the violence and duration of the disease will be modified and frequently cut short by the early expulsion of the morbid accumulations. These acrid and irritating matters are the direct and perpetuating cause of the febrile excitement. The fever so induced and perpetuated hastens the metamorphosis of the interstitial tissues of the body, and thus is the labor of depuration augmented, and the liability to local congestions increased. By the long retention of the metamorphosed animal tissues is engendered a peculiarly acrid and corrosive condition of the fluid menstrua, which hold these matters in solution, even to such an extent that they will react upon and destroy the very apparatus in which they circulate. This we have evidence of in the advanced stages of the disease, when an uncontrollable diarrhea sets in, and which is the result of an absolute erosion of the glandular structure of the intestines by their own secretions, which now are of a decided septic character. Thus what was at first but *functional* has become *organic*, and the integrity of the secreting vessels is destroyed by their own legitimate contents. How important, then, that these facts be taken into consideration early in the history of the disease. No matter what may have been the primary causes by which a retention of the waste matters of the system was induced, their retention and consequent accumulation constitutes a morbid condition, an effect, which, remaining uncorrected, becomes a cause or antecedent to the production of further results. For the purpose of meeting the indications, no better agent comes within the province of the healing art than the Podophyllin. We by no means advocate it as a specific, but as being appropriate and reliable in by far the largest majority of cases. In order, however, to be successful with this agent, the conditions heretofore mentioned as governing its employment must be strictly observed.

Podophyllin, in our opinion, is eminently superior to all other remedies as a *resolvent* and *alterative*. In this opinion we are not alone. It is, for this reason, more frequently indicated in the treatment of chronic disease than any other remedial agent. In all disorders of the liver, no matter what their type, we have need to avail ourselves of the curative powers of Podophyllin. Be that organ indolent from any cause, excepting only a deficiency in the blood of the elementary constituents of bile, we have, in the Podophyllin, a safe and certain agent for restoring its functional energy. In this case it proves directly stimulant to that organ, and is instrumental in restoring lost action. If, on the other hand, the condition be one of abnormal excitement, as in diarrhea, dysentery, etc., Podophyllin is equally efficient in regulating the secretive action of that organ. No matter to which side the scale may be turned, Podophyllin may be relied upon to restore and harmonise the functions of secretion. Our views in relation to the peculiar property whereby diverse derangements are regulated by one and the same remedial agent, are more fully set forth under the head of Senecin. Transfer the exposition there given to the Podophyllin, and the phenomenon is explained.

As a derivative, in the discussion and diffusion of local inflammations and congestions, the Podophyllin is, perhaps, without an equal. In the treatment of chronic inflammation of the bladder, we have frequently had occasion to put its peculiar virtues in this respect to the proof, and never have we been disappointed. As a radical means in the cure of this complaint, our success with it has been such that we deem it indispensable. We usually exhibit it in full cathartic doses at bed time, and repeat every second or third night until the more violent symptoms are subdued. It answers well combined with Asclepin in these cases. The auxiliary remedies will consist of mucilaginous and cooling diuretics as a decoction of *Marsh Mallows, Pumpkin-seeds,* or *Cleavers* infused in cold water. Populin, Lupulin, and Hydrastin will also be found serviceable. If calculous deposits are suspected, *borate of soda,* in doses of TWO grains twice a day.

In the treatment of felons, and local inflammations generally, we almost invariably employ the Podophyllin in full doses, and have always found it efficient in modifying the inflammatory action, and abating the violence of the local congestion. As a revellent, we give it the preference over all other remedies.

As an alterative, in the treatment of syphilitic infections, its sanative influences are more certain and reliable than those of mercury, and its operation entirely devoid of any secondary deleterious effects whatever. Not only is this true in regard to primary syphilis, but also of the secondary and tertiary forms, and he who fails with this remedy, when judiciously employed, need not resort to mercurials with any hope of success. It does not cure by changing the type of the disease, inducing a Podophyllo-syphilitic complication, but by eradicating the virus effectually from the system. And when primary syphilis is properly treated with Podophyllin, in connection with suitable auxiliaries derived from the organic materia medica, secondary and tertiary symptoms will seldom or never appear. At least we have never known such a result, and our experience has not been limited. It is a great mistake to suppose the vegetable kingdom incapable of affording a remedy of equal efficacy with mercury in resolving deposits of inflammatory exudations, for in Podophyllin we have that remedy. Whether they arise from pleural or other adhesions of the serous membranes, or from syphilitic or other infections, the Podophyllin will answer an equally good purpose. In these cases it should be given in small doses, say from ONE-EIGHTH to ONE-HALF of one grain, and continued for a length of time, occasionally administering a full dose, if the bowels are not sufficiently relaxed, in order to guard against intestinal accumulations. The best adjunctive in these cases is Asclepin

Piles, when dependent upon a sluggish condition of the

portal circulation, are promptly and radically relieved by the use of Podophyllin. The proper method of using it in this complaint is to commence with a dose sufficiently large to impress the liver thoroughly, and then follow with small doses in combination with Hydrastin, as follows:

℞,
        Podophyllin............ grs.iii
        Hydrastin ........... grs.xxiv.

Mix, and divide into twelve powders. Dose—one, twice or thrice a day, according to the solubility of the bowels. We prefer, however, alternating the Podophyllin with Hydrastin, exhibiting the latter during the day and the former at bed-time. In all cases of chronic disease, the Podophyllin will operate better if given at bed time, as the stomach is then, or should be, free from all other matters requiring digestive action, and can devote its energies exclusively to the appropriation of the medicine.

For jaundice, the Podophyllin should be alternated with Leptandrin, Juglandin, Hydrastin, etc. These should be given in appropriate doses two or three times per day, and a cathartic dose of Podophyllin administered every third or fourth night.

We seldom employ any other medicine than Podophyllin in the treatment of croup, when called to treat the disease in its incipient stages. Our first care is to apply the cold water bandage to the throat, and to have the feet frequently bathed in warm alkaline water. We then exhibit a full dose of Podophyllin, combining it as circumstances require, and seldom find occasion for other medicine, or even a repetition of the same. If other medicine be absolutely necessary, we employ the Asclepin in solution, with, occasionally, a few drops of the Wine Tincture of Lobelia. If this plan of treatment be adopted early, it will seldom disappoint the practitioner. The peculiar alterative and resolvent properties of the Podophyllin render it invaluable in arresting the progress of membranous croup.

In chronic constipation of the bowels, arising from hepatic

torpor, we know of no remedy more to be relied upon than the Podophyllin. To ensure success, the remedy must be persevered in. In one case of fifteen years' duration, we continued the use of this medicine for one year, exhibiting it on an average every alternate night, and with complete success. Tonics should be used in connection.

For scrofula, ophthalmia, otorrhea, eruptions of the skin, and for all diseases arising from, or dependent upon, tardy depuration, hepatic aberation, local obstructions, defective secretion, or a vitiated condition of the blood and fluids from any cause, Podophyllin is the radical remedy. It arouses the latent energies of the system, and paves the way for further medication. Podophyllin exercises a remarkable control over the sanguiferous system, removing capillary obstructions, and equalising the circulation. The exhibition of a dose of Podophyllin is frequently followed by a decided increase of temperature on the part of the skin, and patients sometimes imagine that the medicine is going to induce a fever. Many who have been troubled with unequal circulation and coldness of the extremities for months, are permanently relieved by a single dose. In apoplexy, as soon as the patient is restored to consciousness, we generally exhibit a full dose of Podophyllin and Leptandrin, and the early employment of the same prescription will generally prevent an attack, when taken on the approach of the premonitory symptoms.

In cholera morbus, as soon as the vomiting and spasms are allayed, we invariably exhibit the same combination, adding to it, if occasion requires, Dioscorein, or Caulophyllin, or Asclepin, etc. If the affection be accompanied with hepatic congestion, it will relieve the pain in a very short time, and prove the very best anodyne that can possibly be used.

For the convulsions of dentition, we give the Podophyllin preference over all other remedies. While others administer antispasmodics, anodynes, etc., we give Podophyllin, and we have never been disappointed in our expectations. The fact is, in all congestions of the hepatic system, Podophyllin is without an equal as an anti-spasmodic. Hence, as soon as the difficulty is determined to arise from biliary obstruction, palliative means should be dispensed with, and the radical remedy, Podophyllin, immediately exhibited. The timely use of Podophyllin during the period of dentition will obviate all liability to convulsions. As acidity of the stomach predominates during this period, means must be employed to neutralise it. Lime water, in doses of a teaspoonful three or four times a day, is the best remedy we are acquainted with. If this precaution has been neglected, the Podophyllin may be combined with supercarbonate of soda, when exhibited, otherwise it may fail of its effect. It will be remembered that we have stated that the operation of Podophyllin is negatived by the presence of lactic acid. Bearing in mind the fact, also, that the food of children at this age consists chiefly of milk, the most ready source for the production of lactic acid, the necessity of our admonition will be apparent. If the symptoms indicate the presence of acrid ingesta in the stomach, an emetic of the Wine Tincture of Lobelia should precede the exhibition of Podophyllin, as more prompt relief will thereby be afforded. We have attended many cases of so-called congestion of the brain in infants, which we have demonstrated to have arisen from the presence of a considerable quantity of acrid ingesta in the alimentary canal. In one case, a child of eight months old, we removed, by means of Lobelia and Podophyllin, *one and a half pints of solid casein.* This matter so expelled was in a high state of putrefactive fermentation. The child was being reared by hand, as it is called, that is, fed upon cows' milk. The expulsion of these morbid accumulations was followed by an abatement of all the symptoms, rendering other medication, except a little Asclepin and Con. Tinc. Veratrum, to soothe the excited nervous and sanguiferous systems, unnecessary. We mention this case as simply illustrative of many that have occurred under our observation, both in our own practice and that of others, and to demonstrate the efficacy, reliability, and safety of the remedials above mentioned, in the treatment of the diseases to which infants are liable.

But of all the valuable properties pertaining to the Podophyllin, perhaps none are more remarkable than its power, in connection with olive oil, of removing biliary concretions. That it does possess this power we have demonstrated again and again. The symptoms indicating the existence of these concretions are manifold, yet so well marked, that the diagnosis is not difficult. The ordinary symptoms indicating a functional disturbance of the liver, such as furred tongue, bad taste in the mouth, sallowness of the skin, eyes, etc., are usually present. The special symptoms are, in many cases, a seated pain in the right epigastrium, which both internal and external means fail to alleviate; a feeling of distension or fulness in the region of the liver; bowels sometimes constipated, at other times diarrhea; but the most certain symptom is alternate diarrhea and constipation; severe pain in the head, accompanied with nausea and vomiting of bilious matter; sometimes the patient is attacked at night with a severe spasmodic pain in the region of the liver, with difficult respiration, and is only relieved by free vomiting, which requires to be artificially produced, as the spasm is generally so great that it will not take place spontaneously. Other symptoms, as emaciation, extreme sallowness of the skin, cough, sudden faintings, scanty and high-colored urine, etc., are sometimes present. Many cases of periodical sick headache are entirely dependent upon this cause.

The proper plan of treatment in these cases is as follows: exhibit at bed time the following powder:

℞.
  Podophyllin...................... grs.ij
  Euphorbin...................... grs.j
  Caulophyllin ................... grs.ij.

Mix. It is best taken in a little water. The next morning, as soon as the nausea attending the operation of the powder has subsided, administer EIGHT OUNCES of pure Olive Oil. This quantity is the average dose for an adult. We have known as high as SIXTEEN ounces to be given at a single dose, but the patient was of unusual physical development. We have frequently administered SIX and EIGHT ounces to females, and never without the most beneficial results. The oil will sometimes produce a considerable degree of nausea, and usually operates upon the bowels in the course of four hours. We have known as many as two hundred of these concretions, varying in size from that of a small pea to that of a hazel nut, to be passed after the administration of a single dose of the Podophyllin and oil. It is always advisable to administer half the quantity of oil on the second morning, as some of the concretions are liable to lodge in the bowels, giving rise to flatulence, pain and irritation. If there is reason to suspect that more of the concretions remain unexpelled, the same course of treatment must be repeated. It is requisite, in all cases, to give the Podophyllin in sufficient doses to relax the hepatic constriction thoroughly previous to exhibiting the oil. The dose above advised will of course require to be increased or diminished according to circumstances. A smaller quantity of oil may also sometimes answer the purpose, although less than four ounces will prove inefficient. We seldom give less than six. The combination of the Podophyllin may also be varied at the option of the practitioner. Leptandrin may be substituted for the Euphorbin, and Gelsemin for the Caulophyllin. We give preference to the Euphorbin, however, having met with better success in the use of that agent as an adjunctive

After the operation of the above medicine, the patient should be placed upon a laxative and tonic course of treatment. The following we have found excellent:

℞.
    Leptandrin  · · · · · ·  ℨj.
    Cornin  · · · · · ·  ℨij.

Mix. Dose—three to five grains three times a day. Hydrastin will answer a good purpose in some cases. Juglandin is also excellent, and may be combined with Cornin in equal proportions.

In the relief of suppression and retention of urine, we have found the Podophyllin of exceeding utility, as a radical remedy. We remember one case in which the catheter had been used, on an average, twelve times in twenty-four hours, for four weeks, and which was promptly and permanently relieved by a single dose of Podophyllin, rendering the further use of the catheter unnecessary. In all derangements of the urinary apparatus, Podophyllin will be found one of the best alterative diuretics that can possibly be employed. It operates not so much by increasing the flow of urine, as by restoring the secreting power of the kidneys. It is very effectual in removing uric acid deposits, and corrects the diathesis giving rise to the superabundant formation of that constituent of the urine. It is of exceeding utility in all calculous affections, by reason of its peculiar alterative, resolvent, and diuretic properties. Frequently, during its operation, considerable pain will be felt in the region of the kidneys, followed by a flow of urine highly charged with calculous sediment.

In the treatment of the various types of skin diseases, we have, in the Podophyllin, one of the best constitutional remedies that can possibly be employed. It exercises a peculiar influence over the sub-cutaneous glandular system, which, in fact, is but typical of its influence over the entire glandular structure of the system. Its action upon the animal economy is universal, not a gland or tissue escaping its sanative impress. It awakens power when latent, quickens the functions when tardy, resolves viscid deposits, restores and harmonizes the functions of secretion, removes obstructions, promotes depuration, dissipates capillary congestion, equalizes the circulation, and restores equilibrium of action to the nervous system. All this it does without corroding the tissues, or vitiating the fluids, promoting the expulsion of nothing but that which has become effete, entering nto no abnormal combinations, nor in any wise impairing the integrity of the materials of organic formation. Having expended its therapeutic powers upon the various functions of the system, it is itself depurated along with other waste matters, leaving none other than physiological traces of its former presence. We are aware that our language is laudatory in the extreme, but we have no hesitancy in expressing our convictions upon a subject so pregnant with the best interests of suffering humanity. The truth of our expressions has been fully and repeatedly demonstrated by hundreds of the most intelligent of our profession, hence we stand not alone in our estimation of the remedial agent under consideration.

Of the special combinations of Podophyllin, we shall not have much to say in the present article. They are noticed throughout the work in connection with other agents. We shall, therefore, leave it to the judgment of the practitioner to form such combinations as his judgment may dictate. We would again state that Podophyllin will take its own time to operate, independently of the agent with which it may be combined. The average time required for the operation of Podophyllin is eight hours. The variations in this respect will depend in a great measure upon the readiness with which it is dissolved and absorbed. Hence any means by which those processes may be facilitated will tend to quicken its operation. By triturating it thoroughly with Asclepin, it will the more readily pass into solution, and in this form is appropriate in the treatment of skin diseases, pleural adhesions, capillary congestions, bilious and typhoid fever, dysentery, iritis, etc. With Baptisin for amenorrhea and defective menstruation. Triturated with gum arabic, one part in eight, it answers an excellent purpose in disorders of the bowels. We wish it distinctly borne in mind, in order to ensure success, that it is not sufficient, in the use of this remedy in the treatment of violent attacks of acute disease, as well as in chronic cases, to simply produce a cathartic effect upon the bowels, but the action must be promoted until the evacuations show that the morbid accumulations are expelled, and the secretions evince a more healthy appearance, Podophyllin is sometimes very tardy in its operation, not acting under eighteen or twenty hours, and frequently it will operate more freely during the second twenty-four hours than during the first. In cases of chronic disorders of the liver, spleen, and other viscera, considerable pain will frequently be experienced in the diseased organ during the operation of the medicine. Sometimes the pain will be in the liver, at others in the spleen, again in the kidneys, in the back of the neck and head, in the pleura, intercostales, etc., but these symptoms will subside with the operation of the medicine, and are favorable indications, showing that the remedy is at work arousing the dormant energies of the system.

The average cathartic dose of Podophyllin is TWO GRAINS. An emeto-cathartic dose, from THREE to FIVE GRAINS, although ONE or TWO grains will frequently vomit. As an alterative, from ONE-EIGHTH to ONE-HALF of ONE grain. In combining it with other agents of similar properties, the quantities admit of some reduction. If much pain or griping is experienced during its operation, it may generally be readily relieved by administering freely of warm ginger tea. When, however, the pain is obstinately seated in the small intestines, it will be relieved only by a motion of the bowels, and upon observing the character of the discharge, it will be found to consist, in most cases, of a clear, jelly-like substance, plastic lymph, which is the material of which the false membrane that frequently lines the small intestines is formed. The pain will abate as soon as this matter is expelled.

In case Podophyllin be taken by mistake, or in over-doses, it is readily counteracted by lactic acid, the most ready source of which is sour milk, or buttermilk, which should be freely administered.

# MYRICIN.

Derived from *Myrica Cerifera.*
Nat. Ord.—*Myricaceæ.*
Sex. Syst.—*Diœcia Tetrandria.*
Common Names.—*Bayberry, Tallow Berry, Wax Myrtle, Wax Berry, Candle Berry, etc.*
Part Used.—*Bark of the Root.*
No. of Principles.—*Two,* viz., *resinoid* and *tannin.*
Properties.—*Alterative, astringent, stimulant, diuretic, antispasmodic, and anti-syphilitic.*
Employment.—*Apthous affections, scrofula, diarrhea, dysentery, jaundice, leucorrhea, catarrh, polypus, fistula, suppression of urine,* to allay *false labor pains, burn, chancres* and whenever a *stimulating astringent* is needed.

FEW of the simple agents of the materia medica are of more practical and frequent utility than the Myricin. We class it among the simpler agencies, because, while it possesses specific and decided therapeutic properties, it is entirely innoxious in itself. With the exception of a mild diuretic action, it is never visibly evacuant, except in very large doses, when it sometimes proves emetic.

Amongst the affections in which the Myricin has been found peculiarly serviceable, we may first mention apthous affections of the mucous surfaces. It is valuable both as a local and as a constitutional remedy. In the various forms of stomatitis, ulcerative sore throat, nursing sore mouth, and ulcerations of the stomach and bowels, it has proved of great utility. The average dose for internal administration is THREE grains, which, in bad cases, may be repeated every three hours. Locally, it may be used in the form of a gargle, ONE DRACHM to HALF a pint of boiling water. It is usual to combine it with other astringents, Rhusin being the best for the purpose. They may be combined in equal proportions. If desirable to avoid constipation, it should be alternated with Leptandrin, Juglandin, etc. In painful ulcerative affections of the stomach and bowels, it may be advantageously combined with Lupulin, equal parts, and from THREE to FIVE grains exhibited once in three hours, in warm water.

Myricin has obtained considerable repute in the treatment of scrofula. It is an efficient alterative, and its peculiar stimulant properties are exceedingly appropriate in the cold and languid conditions characteristic of that disease. It should be given in doses of from THREE to FIVE grains three times per day. It is necessary to obviate its astringent effects when used as an alterative, for which purpose it may be combined with such laxatives as the judgment of the practitioner may dictate at the time, although we much prefer alternation. We consider it better practice to employ the Myricin alone during the day, and to exhibit a suitable dose of Podophyllin, Leptandrin, or other cathartic or laxative at bed time. Externally, the Myricin is applied to scrofulous ulcers, the surfaces of which may be sprinkled over with it, or it may be applied by means of a poultice. In the same manner it is an excellent stimulant to old and indolent ulcers. In solution, it is employed as an injection in scrofulous, mammary, and other abscesses.

In diarrhea and dysentery, Myricin is employed with great benefit, but not until the morbid accumulations have been expelled and the functions of the liver regulated. It may then be administered in doses of TWO grains every one to three hours, until the discharges are controlled. In these complaints it may be joined to the Geranin, or Rhusin, etc. To increase its stimulant properties, with Xanthoxylin. In the diarrhea of phthisis pulmonalis, and when the system has been exhausted by profuse colliquitive discharges, with Fraserin, as follows:

℞.
Myricin.............................
Fraserin ........................... aa. ℈j

Mix, and divide into ten powders. Dose—One every two to four hours.

Myricin has been found serviceable in jaundice, in which complaint it may be combined with Apocynin, or Leptandrin, or Euonymin, etc. Enough of the adjunctive agent should be added to overcome the astringency of the Myricin.

Myricin is much employed in leucorrhea, though mostly as a local remedy. ONE drachm may be infused in a pint of boiling water, and used in suitable quantities as an enema. For the relief of fetid leucorrheal discharges, ONE drachm each, of Myricin and Baptisin should be infused in a pint of boiling water, and used as an injection, alternated with a solution of chloride of lime, ONE ounce to a quart of cold water.

Myricin, used as a snuff, will relieve catarrh, and has been found beneficial in some forms of nasal polypus. In the latter affection, it may be rendered more efficient by combining it with Sanguinarin.

In solution, Myricin is employed as an injection, to promote the healing of fistulous openings after they have been converted into simple ulcers by the use of suitable remedies. It will be found reliable for this purpose when the parts are tardy in healing.

We have found the Myricin effectual in relieving suppression of the urine, for which purpose we usually employ it in the form of an enema. From ONE-HALF to ONE drachm may be administered in SIX ounces of warm water, the patient retaining it as long as possible. If the first should not be retained a sufficient length of time, repeat until the desired effect is produced. To render it more effectual, from ONE-HALF to ONE ounce of the Wine Tincture of Lobelia may be added to each injection. The same will be found admirable for relieving the pain and promoting the expulsion of renal calculi. While using the above, the Myricin may be administered internally, in doses of FIVE grains every two hours, in warm water. To add to its efficacy it may be joined with Populin.

But perhaps the most remarkable feature of the Myricin is its power, in connection with Lobelia, of allaying false labor pains. The peculiar therapeutic property here manifested is the result of the combination. Neither will answer the purpose alone. As soon as the pains are ascertained to be spasmodic, place the patient in bed, and administer the following:

℞.
 Myricin   -   -   -   -   -   grs. xv.
 Wine Tinc. Lobelia   -   -   -   ℨ ss.
 Boiling Water   -   -   -   -   ℨ j.

Add the Myricin to the boiling water, and after a few minutes the Tinc. Lobelia. Exhibit at one dose, and repeat in two hours, if necessary. This will seldom or never disappoint the practitioner, and rarely is a second dose necessary. It allays the pains, quiets the nervous system, and postpones parturition to the proper period. Delivery will frequently be delayed from one to four weeks, and the matured energies of the system will then ensure a safe and easy accouchment.

Myricin is an excellent application to burns after the pain and inflammation has measurably subsided. Applied in time, it heals them without suppuration. For this purpose it is best dissolved in alcohol, from TWO to FOUR drachms to the pint. Apply cloths wetted with the tincture.

In the treatment of mild chancres, the Myricin will be found efficient as a local application in a majority of cases. Fill the sore with the dry Myricin, and dress with cold water. Renew twice or thrice a day. Internally, Myricin is of great utility in the treatment of syphilitic infections, possessing considerable power in itself as an anti-syphilitic, as well as promoting the action of other alteratives. In this disease it should be given in doses of FIVE grains three times a day, and persevered in for a length of time, alternated with an occasional dose of Podophyllin. In all languid and cankered conditions of the stomach and bowels, the Myricin is admirably calculated to arouse the latent forces of the system, detach false membranous formations, and promote the action of auxiliary remedies. To prepare the stomach, and facilitate the operation of emetics, there is nothing better than Myricin. Administer in plenty of warm water. In cases of atony of the digestive apparatus and general debility, the Myricin will be found one of the most serviceable agents in the range of the materia medica.

# EUONYMIN.

Derived from *Euonymus Americanus.*
Nat. Ord.—*Celastraceæ.*
Sex. Syst.—*Pentandria Monogynia.*
Common Names— *Wahoo, Burning Bush, Spindle Tree, Indian Arrow Wood, etc.*
Part Used—*The Bark.*
No. of Principles—*three,* viz., *Resinoid, neutral, and alkaloid.*
Properties—*Tonic, laxative, alterative, and expectorant.*
Employment—*Dyspepsia, constipation, dropsy, hepatic torpor, and affections of the respiratory system.*

IN medicinal doses, Euonymin is laxative, tonic, **alterative,** expectorant, and feebly diuretic. It is also accredited with a degree of anti-periodic power. In very large doses it proves a drastic cathartic, its operation being attended with a death-like nausea, excessive tormina, prostration, and cold sweats. The dejections from the bowels are violent, profuse, and accompanied with much flatus. From these symptoms, however, the patient soon recovers.

We esteem the Euonymin a remedy of great value. In the treatment of indigestion arising from hepatic torpor, it will be found of excellent service. It is powerfully tonic, and while it deterges and resolves viscid deposits, and promotes the various secretions, it imparts a decided and permanent tone to the various functions. The average dose of the Euonymin is TWO grains. This quantity may be given twice or thrice a day as occasion requires. It may be joined with other tonics when desired, as the Cornin, Hydrastin, Fraserin, etc., or with antispasmodics and nervines, as the Cypripedin, Caulophyllin, Lupulin, Scutellarin, etc. When a stimulant is needed, with Xanthoxylin, and in some cases of scrofula, torpor of the lacteals, and syphilitic diseases, with Myricin.

For the relief of obstinate constipation of the bowels, the Euonymin is one of the most reliable agents we possess. It is not as prompt as many other laxatives in its operation, some two or three days frequently elapsing before it manifests any effect upon the system, but it makes amends for its tardiness by the permanency of its influence. In order to effect a radical cure, the Euonymin must be persevered with, in moderate doses, for a considerable length of time. It operates slowly but surely.

Euonymin has been found useful in the treatment of dropsy, in which complaint it proves efficacious by reason of its resolvent, diuretic, and tonic properties. Its diuretic influence is more secondary than primary, being the result of increased absorption. It is of great utility in dropsy, after the effusions have been removed, for the purpose of toning up the system and preventing a return. This it accomplishes by maintaining the integrity of the secretive action of the system. Although not, in the proper sense, a diaphoretic, it promotes the depurative action of the skin, and this, together with its laxative power, renders it valuable for the prevention and removal of serous exudations.

In the treatment of hepatic torpor, we have, in the Euonymin, a remedy deserving of much confidence. It may be combined with any other agent or agents that the judgment of the practitioner may deem indicated, or may be alternated with such auxiliaries as the necessity of the case demands. We prefer the latter course. In indigestion arising from hepatic torpor, and accompanied with acidity, the Juglandin will be found an excellent adjunctive, of which two parts may be combined with one of Euonymin.

Euonymin is an excellent remedy in affections of the respiratory apparatus, as bronchitis, laryingitis, coughs, colds, influenza, and incipient phthisis. Asthma arising from a disordered action of the liver may be most effectually cured by means of the Euonymin. In pneumonia, as soon as the inflammatory symptoms are subdued, the Euonymin operates admirably as an expectorant, promoting at the same time the depurative action of the skin, kidneys, and bowels, thus relieving the lungs by promoting the expulsion of effete matter through the proper channels, and imparting tone to the digestive and assimilative apparatus. Hectic fever is frequently arrested by means of the Euonymin, and chronic cases of intermittent fever have been cured by a persevering use of the same remedy, thus seeming to entitle it to the appellation of anti-periodic. In the treatment of coughs, colds, and influenza, it is better to give the Euonymin in **small** and oft-repeated doses, say HALF a grain every two hours. The same course answers well in pneumonia. In the treatment of the form of asthma above mentioned, the use of the Euonymin should be preceded by a thorough dose of Podophyllin.

As a laxative and tonic, from TWO to FOUR grains may be given. As an expectorant, from ONE-FOURTH to ONE grain. In chronic disease, **the system should first be cleansed with Podophyllin.**

## CONCENT. TINCTURE EUONYMUS AMERICANUS.

Derivation, properties, and employment, same as the Euonymin. Contains all the virtues of the bark in a concentrated and reliable form. Average dose, FOUR drops. Convenient for adding to mixtures, and for combining with other of the concentrated tinctures. Said to be of some efficacy as a vermifuge, for which purpose it may be joined with the Con. Tinc. Chelone Glabra, or Apocynum Cannabinum. It will prove a desirable adjunctive, on account of its laxative and tonic properties. Combined with the Con. Tinc. Xanthoxylum Frax., will be found useful in torpor of the lacteals. Convenient and useful as an expectorant, in coughs, colds, influenza, asthma, phthisis, pleuritis, pneumonia, etc., in doses of ONE drop every hour or two, as may be necessary.

# OIL OF ERIGERON.

Derived from *Erigeron Canadense.*
Nat. Ord.—*Asteraceæ.*
Sex. Syst.—*Synyenesia Superflua.*
Common Names.—*Fire Weed, Canada Fleabane, Colt's Tail, Scabious,* etc.
Part Used.—*The Plant.*
Properties.—*Astringent, styptic, and diuretic.*
Employment.—*Uterine hemorrhage, hemoptysis, hematamesis, hematuria, menorrhagia, dysmenorrhea, uterine leucorrhea, gonorrhea, gravel,* and other affections of the *urinary apparatus.* Locally, in *rheumatic affections, enlargement of the tonsils, neuralgia, spinal irritation,* etc.

The Oil of Erigeron is, in our estimation, the most valuable remedy of its class. Although not a specific, it is undoubtedly the best agent we possess for the relief of uterine hemorrhage. The dose of the oil in these cases is from FIVE to TEN drops, repeated once in from thirty to sixty minutes, according to the urgency of the symptoms. It will act more promptly, being rendered more diffusible, by being previously dissolved in alcohol. In addition to internal adninistration, it may also be applied locally with the best results. A case occurred under the observation of the writer over twenty years ago, in which the patient, from excessive loss of blood, was reduced to a comatose condition, and incapable of swallowing. A piece of cotton wool, saturated with the oil, was introduced into the vagina and placed in close juxtaposition with the mouth of the uterus, when an instantaneous stop was put to the bloody flow. The patient is still living, in good health, having attained the age of sixty-three years. During the past season we were consulted in a similar case, in which we advised the adoption of the above plan, and with complete success.

Auxiliary agents may be employed in connection with the oil, if deemed advisable. The Myricin, Lycopin, Trilliin, Geranin and Hamamelin are all good, and may be given in suitable doses in warm water. An infusion of Avens root, *Geum Rivale*, answers an excellent purpose. In passive hemorrhages, Cerasein, or the Oil of Capsicum, will answer the best purpose. ONE drop of the Oil of Capsicum should be given with each dose of the Erigeron.

For hemoptysis, we alternate the oil with Lycopin. If the condition of the stomach does not contra-indicate, we use the oil in the following manner:

℞.
  Oil Erigeron Canad. ............... gtt.xv
  White Sugar ...................... ℨij
  Water ............................ ℥ij

Triturate the oil thoroughly with the sugar, and add the water. If sufficient care be exercised, the oil will be completely suspended in the water. If the hemorrhage is severe, give one teaspoonful every ten or fifteen minutes, until it is arrested, and then at intervals of from two to four hours. As soon as the urgent symptoms are allayed, in order to effect a radical cure, alternate with Lycopin, giving a dose of the oil morning and evening, and from TWO to FIVE grains of the Lycopin at noon and at bed time. Or better, make a solution of the Lycopin, fifteen grains to FOUR ounces of warm water, and let the patient take a tablespoonful once in three hours. If diaphoretics are needed, combine the Lycopin with Asclepin. To obviate the astringent effects upon the bowels, Leptandrin, Euonymin, Hydrastin, Podophyllin, Menispermin, etc., may be employed.

In the treatment of hematamesis, small doses of the oil frequently repeated, will answer a better purpose. Prepared as above directed, we employ it in this affection, and in hematuria, menorrhagia, and dysmenorrhea. In the latter two complaints we have made much use of it, and with the most gratifying success. It allays the spasmodic pains accompanying dysmenorrhea, and restrains, without suppressing the menstrual flow, when too profuse. One teaspoonful of the above preparation of the oil may be given every one, two, or three hours, according to the urgency of the symptoms. When gastric derangement forbids the use of sugar, the oil may be dissolved in alcohol and mixed with water, or exhibited in mucilage of gum arabic, or slippery elm. Or it may be formed into pills with bread, or any other suitable excipient.

Oil of Erigeron exercises considerable control over the heart and arterial system, acting as a sedative. We have found it serviceable in allaying palpitation of the heart, particularly when arising from uterine irritation. From TWO to FIVE drops may be administered at a time, and repeated as occasion requires. The remarkable sanative influences exercised by this agent on the uterine system, gives it a wide range of employment. In combination with Oil of Stillingia, we have used it with remarkable success in the relief of those peculiar headaches accompanying defective menstruation.

℞.
    Oil Erigeron - - - - -
    " Stillingia - - - . aa. ℨj.
Mix. Dose—TWO drops, THREE times per day. This has answered our purpose when other remedies failed. The same combination will be found of service in uterine leucorrhea, and in gonorrhea. We have used the Oil of Erigeron alone in gonorrhea, with the most marked and beneficial results. It may be added to the mixtures used in that complaint,

although we prefer to administer it alone, usually giving it twice a day, in the morning and at bed time. It allays the scalding of the urine, and assists materially in cutting short the disease. It is of much service in inflammation of the kidneys and bladder, and in gravelly affections. It harmonises and gives tone to the functions of both the uterine and urinary apparatus. Its diuretic power consists more in an alterative property, regulating rather than increasing the secretion of urine.

Locally, we have used the Oil of Erigeron in a variety of complaints, and with the most beneficial results. As an application to inflamed and enlarged tonsils, and inflammation and ulceration of the throat generally, this remedy has few superiors. For the purpose of applying to the tonsils, it should be dissolved in alcohol, in the proportion of ONE drachm of the oil to from ONE to TWO ounces of alcohol. Apply with a probang two or three times a day. We also apply it to the throat, externally, at the same time, for which purpose we dissolve ONE ounce of the oil in from EIGHT to SIXTEEN ounces of alcohol, according to the degree of stimulation desired. Bathe the throat freely several times a day, or wet a cloth in the tincture and bind on the parts. If there is much swelling and inflammation, over the cloth so wetted apply the cold water bandage. This application will produce a burning sensation of the skin, much resembling that produced by Capsicum, but will not vesicate. This liniment will also be found excellent as an application to other local inflammations, as painful tumors, rheumatic swellings, spinal irritation, chilblains, etc. We have frequently applied the pure oil with excellent effect in sciatica, neuralgia, rheumatism, etc. It is powerfully rubefacient, but we never remember to have seen it vesicate. We mention this fact, as we have seen it stated by some writers that it is too acrid for topical use.

In syphilitic ulcerations of the throat, after the use of proper caustics, we know of no better application for allaying the inflammation and promoting the healing of the ulcers. For this purpose, ONE part of the Oil should be dissolved in from

FOUR to EIGHT of alcohol. The same will be found of service as an application to indolent ulcers, and certain forms of cutaneous eruptions.

The Oil dissolved in alcohol, ONE drachm of the former to TWO ounces of the latter, has been found serviceable for the purposes of inhalation in hemoptysis and other affections of the respiratory organs. ONE drachm of the above tincture, added to one pint of water, and evaporated in a suitable vessel, will answer for several inhalations. It is excellent where there is a tendency to hemorrhage, and where the air surfaces are extremely susceptible to the differences in temperature of the air inhaled. In the latter stages of phthisis, and in pneumonia, asthma etc., much benefit will be derived from this inhalation. It stimulates secretion, while it relaxes and soothes the nerves.

# ALNUIN.

Derived from *Alnus Rubra*, (*A. Serrulata of Willdenow.*)
Nat. Ord.—*Betulaceæ.*
Sex. Syst.—*Monœcia Tetrandria.*
Common Names.—*Tag Alder, Swamp Alder, etc.*
Part Used.—*The Bark.*
No. of Principles.—*three*, viz., *resin, resinoid,* and *neutral*
Properties.—*Alterative, resolvent, tonic* and *sub-astringent*
Employment.—*Scrofula, eruptions of the skin, rheumatism, syphilis,* and whenever an *alterative* is required.

The **Alnuin** is chiefly valuable as an alterative, resolvent, and tonic, its astringent properties being but feeble, and in no wise interfering with its properties as an alterative. We have been familiar with the employment of the Alnus and its preparations for many years, and our experience enables us to speak in very decided terms as regards its therapeutic value. We esteem it one of the best simple alteratives and resolvents possible to be employed in scrofula, cutaneous eruptions, and in all affections arising from a vitiated condition of the blood and fluids. In order to reap the full value of the Alnuin, its use must be persevered in for a considerable length of time, and we deem alternation preferable to combination, when it is desirable to employ auxiliary alteratives. It is slow, but certain in its operation, resolving viscid deposits, promoting secretion and depuration, increasing the appetite, and giving tone to the digestive apparatus. Although not strictly a diuretic, it nevertheless exercises a peculiar alterative influence over the kidneys and urinary apparatus generally, hence is valuable in the treatment of chronic rheumatism, erysipelas, gonorrhea, gleet, syphilis, gravel, catarrh of the bladder, etc. The average dose of the Alnuin is THREE grains, three times per day. In many cases the dose may be advantageously increased to TEN grains. It seldom or never offends the stomach, hence is peculiarly serviceable in the treatment of patients possessed of a very susceptible organisation. It is appropriate and useful in the convalescing stages of acute diseases, as it obviates the plasticity of the secretions, and at the same time promotes the appetite, digestion, and depuration, thus manifesting the powers of a general tonic.

When combinations are desired, they should be made compatible with the existing necessities. Thus, in rheumatism, the Alnuin may be joined with Macrotin, as follows :

℞.
    Alnuin ............................. ℨss.
    Macrotin ........................... grs. v.

Mix and divide into ten powders. Dose—One, three times a day, or with Phytolacin :

℞.
    Alnuin ............................. ℈ij.
    Phytolacin.......................... ℈j.

Mix, and divide into twenty powders. Dose—same as above. In scrofula it may be desirable to join it with more decided tonics. If laxative properties are indicated, with Euonymin or Hydrastin.

℞.
    Alnuin ............................. ℈j
    Euonymin ........................... grs. x.

Mix, and divide into ten powders.

Or,

℞.
  Alnuin .................................
  Hydrastin............................ aa. ℈ ij.
Mix and divide into twenty powders. Dose of either—one powder three times a day. When the simple tonics are indicated, as in the convalescing stages of dysentery, diarrhea, cholera, etc, Fraserin, or Cornin, or Cerasein, will be appropriate. If astringent tonics are required, in order to control a tendency to diarrhea, the Myricin, or Rhusin, or Lycopin, or Trilliin should be employed. In the treatment of scrofulous and indolent ulcers, eruptions of the skin, rheumatism, etc., Xanthoxylin will be found a most valuable adjunctive.

In cases of general debility, particularly of the aged, the Alnuin will be found peculiarly serviceable. While it is not perceptibly evacuant, it nevertheless imparts a healthful impetus to the various functions of the system, proving itself a true constitutional alterative. Of course the dose must be regulated according to the age, sex, and condition of the patient, the chief consideration being to give enough to bring them fully under its influence. When the liver is involved in the existing difficulty, the judicious use of Podophyllin, Leptandrin, Juglandin, etc., will much facilitate the cure; and in all cases, when the liver is primarily deranged, should not only precede, but be occasionally alternated with the Alnuin.

# VIBURNIN.

---

Derived from *Viburnum Opulus.* (*V. Oxycoccus.* Pursh.)
Nat. Ord.—*Caprifoliaceæ.*
Sex. Syst.—*Pentandria Trigynia.*
Common Names.—*High Cranberry, Cramp Bark, etc.*
Part Used.—*The Bark.*
No. of Principles, *four, viz., resinoid, two resins and alkaloid.*

Properties.—*Anti-spasmodic, anti-periodic, expectorant, alterative* and *tonic.*

Employment.—*Cramps, spasms, convulsions, asthma, hysteria, chorea, intermittent fever, pneumonia, dysmenorrhea,* to *prevent abortion,* and to *relieve after-pains.*

The Viburnin is a safe, certain, and reliable anti-spasmodic, for which property it is chiefly valuable. For the relief of cramps and spasmodic pains, no matter from what cause they arise, we know of no remedy of so great general utility. It exercises a wonderful control over muscular fibre, and acts with great promptitude. Although in small doses it is esteemed a tonic, yet we know that in full doses, and continued for a few days, it will most effectually relax the nervous system, and render physical exertion somewhat of a task.

The average dose of the Viburnin is TWO grains, although admitting of being increased to TEN grains with advantage, and of being repeated at intervals of from twenty to sixty minutes until the desired effect is produced. We have used the Viburnin quite extensively, and esteem it an almost indispensible agent of the materia medica. Cramping pains in the limbs, whether arising from the irritation produced by a gravid uterus, or from a fracture of the bone, or in females past the turn of life, and yet troubled with some uterine disturbance, are more generally and radically relieved by the Viburnin than any other remedy, the Gelsemin, perhaps, excepted. For the cramps with which females are afflicted during the period of utero-gestation, it is a safe and certain remedy. For the cramping pains sometimes occurring as sequents to the fractures of bones, we have found it equally efficacious. In asthma and pneumonia, as well as in intermittent fever, it seems of much service, not only correcting the plastic condition of the blood, relaxing or preventing muscular spasm, and acting as an expectorant, but also seeming to manifest considerable anti-periodic power, and so prolonging the remissions, and lessening the tendency to a return. In dysmenorrhea we have used it with the most decidedly beneficial results, both alone and in combination with other agents. For the relief of after pains it is equally beneficial. When abortion is threatened, as the result of over exertion or mental excitement, we have, in the Viburnin, one of the most reliable remedies for its prevention. It allays false labor pains, relaxes spasm, and soothes and harmonises the action of the nervous system. The patient should be brought as quickly as possible under its influence, and perfect quiet enjoined. Notwithstanding its peculiar control over spasm, we have never found it to interfere with true labor pains. We have frequently made use of it during parturition, when the pains were scattered, extending to the thighs and knees, and with the most beneficial results.

Viburnin admits of many combinations, most of which will readily suggest themselves to the practitioner. For dysmenorrhea and after pains, the following is our favorite formula:

℞.
    Viburnin........................
    Caulophyllin .................... aa. ℈j
    Gelsemin ........................ grs. v.

Mix, and divide into ten powders. Dose—one, every two hours, or, in severe cases, every hour, until relieved. This will be found one of the most effective combinations that can possibly be made.

In order to render permanent the good results produced by Viburnin, it is advisable to follow with tonics, as soon as a remission of the symptoms for which it was exhibited occurs. The list embraced in this volume will afford an opportunity for a judicious selection. Quinine, iron, etc., may also be employed at the discretion of the practitioner.

Viburnin has been found remarkably efficacious in relieving the pains accompanying diarrhea, dysentery, and cholera morbus, and also in flatulent and other forms of colic. For use in these complaints it may be joined with Asclepin, or Caulophyllin, or Gelsemin. It will increase the anti-spasmodic power of Dioscorein, and may be joined with it in the treatment of bilious colic. When a tonic is indicated, Fraserin will be found to operate remarkably well in connection with the Viburnin. Finally, as an anti-spasmodic, Viburnin may be relied upon in all cases with confidence, and will seldom disappoint the expectations of the practitioner. It possesses no narcotic property whatever.

# CORNIN.

Derived from *Cornus Florida*.
Nat. Ord.—*Cornaceæ*.
Sex. Syst.—*Tetrandria Monogynia*.
Common Names.—*Dogwood, Boxwood, Flowering Cornel* etc.
No. of Principles—*two*, viz., *resinoid* and *neutral*.
Properties.—*Tonic, stimulant, anti-periodic* and *astringent*
Employment.—*Intermittent* and other *fevers, indigestion, debility*, and the *convalescing stages of many acute diseases.*

As a tonic, the Cornin ranks high in the estimation of all who have employed it. Its anti-periodic power renders it of peculiar value in the treatment of intermittent and other periodic fevers. We have employed it with much success in the cure of fever and ague, either alone, or joined with Macrotin and Xanthoxylin. The average dose of the Cornin is THREE grains, but may be increased to TEN grains in some cases with advantage. The Cornin will be tolerated by the stomach when other tonics are rejected. By many it is esteemed a reliable substitute for quinine, but this opinion, perhaps, needs some qualification. It is certain that Cornin has cured fever and ague when quinine had failed, and that in all cases where the latter cannot be employed, in consequence of a peculiar idiosyncracy, the Cornin answers admirably as a substitute. It is certainly one of the best native substitutes we have for the bark.

When the system is brought under the influence of Cornin, the pulse is accelerated, the temperature of the skin is elevated, and tonicity is imparted to the functions of the system generally. In the treatment of ague and fever, the system should be properly prepared for the influence of tonics by the judicious use of Podophyllin and Leptandrin, and, as soon as a distinct remission occurs, the Cornin then administered in doses of from THREE to FIVE grains every three hours, until the paroxysmal stage is passed, and then continued at longer intervals for three or four days, in order to guard against a return. Acidity of the stomach, if excessive, must be duly neutralised in order to reap the full value of the Cornin. We have frequently used the Cornin in combination with Macrotin and Xanthoxylin, with excellent effect, as follows:

R.
    Cornin ........................ grs.xx
    Xanthoxylin ................... grs. x.
    Macrotin ...................... grs. v.

Mix, and divide into ten powders. Dose—One, every three hours. In quotidian ague, the doses should be repeated every two hours. The quantity of Macrotin must be regulated according to the ability of the patient to bear it. Cornin is most successful in the cure of fevers when the remissions are marked and distinct, hence, if they are obscure, perfect remissions must be induced by the use of Gelsemin, Veratrin, etc., and the Cornin then employed as above directed.

Although Cornin does not possess the power of directly neutralising acidity of the stomach, yet it is of exceeding utility in those cases of indigestion in which that symptom is a troublesome feature. It gives almost immediate relief in that distressing symptom called heart-burn; and its continued use will prove a sure prevention of its recurrence, by restoring the tone of the stomach, and so obviating the tendency to fermentation. Combined with Juglandin, equal parts, it will prove more effective still. From FIVE to TEN grains of the mixture may be taken three times per day. We often advise it to be taken immediately after each meal, as in the case of the Populin, and with the most beneficial results.

In general debility, and in the convalescing stages of acute diseases, the Cornin may be used for all the purposes of a general tonic. Its astringent properties are feeble, and will seldom interfere with its general employment. When a laxative property is needed, we have found it to act admirably in connection with Leptandrin. They may be alternated, using the Cornin during the day, and the Leptandrin at night, or the two may be combined, if desired. The difficulty, in the latter instance, is with the Leptandrin, which, if put up in papers, or in any way exposed to the air, absorbs moisture and hardens. We usually mix the two intimately together, and put them into a tightly corked vial, directing the patient to take as much as will lie upon a three, five, or ten cent piece, as the case may be. True, this is not a very precise way of prescribing, but with medicines so innoxious as these, a grain or two more or less can create no serious disturbance. When preferred, they may be formed into pills.

Cornin has gained considerable repute in the cure of leucorrhea, and, as a general tonic, we have found it of much efficacy in disorders of the female system. In this complaint it may be used in connection with Helonin, or Senecin, or Trilliin, etc. In all cases in which an anti-periodic tonic is indicated, the Cornin may at all times be relied upon as amongst the most efficient of its class.

We desire, in this connection, to direct the attention of the profession to the important difference between the Cornin, of which we have been speaking, and an article of *Cornine* put forth by certain manufacturers, and which is represented as being, "*probably*, a mixture of resin and insoluble alkaloid." A few lines in advance, in the work from which we quote, we are told, in speaking of the Cornus Florida bark, that " water or alcohol extracts its virtues." The wisdom here displayed is, to us, unfathomable. If "water" will "extract its virtues, can the active principle of the bark be 'a mixture of resin and insoluble alkaloid?'" If so, can the water " extract" it, as the alkaloid is represented as being "insoluble," while the resin is equally so, as is demonstrated by the method employed to obtain it, viz., by precipitation of the alcoholic solution by means of water. Not only so, but the "resin" and " insoluble alkaloid" are "mixed," hence more completely "insoluble." The truth is, the active principles of the bark are two in number, consisting of a resinoid and a neutral principle. The latter is the principal and most valuable active constituent of the bark, and is completely soluble in water. This principle it is, in common with that of many other plants, as we have previously had occasion to demonstrate, that incompetent organic chemists throw away with the water from which they have "precipitated" their *probable* active constituents. We see, therefore, that water will extract *a part* of the virtues of the bark only, and that strong alcohol is required to extract the remainder, that is, the resinoid principle. We confess to being somewhat particular upon this point, as the properties and uses of the Cornin, as we have already detailed them, are the result of clinical observation in the use of the two combined principles of the bark, and our reputation as a truthful writer would be jeopardised by applying our remarks **to any "*probable* mixture of resin and insoluble alkaloid."**

# RUMIN.

Derived from *Rumex Crispus*.
Nat Ord.—*Polygonaceæ*.
Sex. Syst.—*Hexandria Tetragynia*.
Common Names.—*Yellow Dock, Sour Dock, etc.*
Part Used.—*The Root*.
No. of Principles—*two*, viz., *resinoid* and *neutral*.
Properties.—*Alterative, resolvent, detergent, anti-scorbutic,* and *mildly astringent and laxative*, much like Rhubarb.
Employment.—*Scrofula, rheumatism, scorbutus, salt rheum, leucorrhea, syphilis, cutaneous eruptions, etc.*

As an alterative, the Rumin is deservedly held in high repute, and is of general and extensive employment in a great variety of diseases. It proves most efficient, however, in scrofula, syphilis, and diseases of the skin. It operates kindly and without excitement, being slow but sure in promoting a healthful action of the depurating functions of the system. Its laxative properties are not displayed, except when given in large doses, and not even then if a considerable degree of hepatic torpor exist. It will be necessary, therefore, to use, in such cases, suitable laxatives in connexion with the Rumin.

When used to an extent sufficient to affect the bowels sensibly it reacts mildly astringent, hence is frequently employed in those cases wherein rhubarb is indicated, as in the asthenic forms of diarrhea and dysentery, and in the diarrhea of phthisis. The average dose of the Rumin is THREE grains, subject to such variations as the circumstances of the case may warrant.

Rumin is seldom employed alone, but generally in connection with other alteratives, or with tonics or laxatives, except in the cases above mentioned. In scrofula it is combined with Ampelopsin, Smilacin, Myricin, Alnuin, etc. In rheumatism, with Macrotin, Sanguinarin, Xanthoxylin, Phytolacin, etc. In scorbutus, with Citrate of Iron, Quinine, Myricin, Oil of Erigeron, etc. In salt rheum, with Stillingin, Leptandrin, Podophyllin, etc., as for all skin diseases. For syphilis, with Corydalin, Ampelopsin, Phytolacin, Smilacin, etc. In leucorrhea, with Helonin, or Trilliin, or Senecin. In short, the suitability of combinations must be determined by the necessities of the case in hand.

# CAULOPHYLLIN.

Derived from *Caulophyllum Thalictroides.*
Nat. Ord—*Berberidaceæ.*
Sex. Syst.—*Hexandria Monogynia*
Common Names.—*Blue Cohosh, Squaw Root,* etc
Part Used.—*The Root.*
No. of Principles.—*Two,* viz., *resinoid* and *neutral.*
Properties.—*Antispasmodic, alterative, tonic, emmenagogue, parturifacient, diaphoretic, diuretic,* and *vermifuge.*
Employment.—*Amenorrhea, dysmenorrhea, menorrhagia, leucorrhea, gonorrhea, to promote delivery, after-pains, dyspepsia, rheumatism, dropsy, hooping cough, hic-cough, hysteria, hysteritis, apthous sore mouth, to expel worms,* etc.

CAULOPHYLLIN is a remedy of frequent and extended utility. Not only is it of almost universal application in the treatment of the diseases peculiar to females, but also in a variety of other affections, both on account of its own remedial properties, and as an agent for modifying the action of other medicines. The average dose of the Caulophyllin is THREE grains. When used for the purposes of an anti-spasmodic, from FIVE to TEN grains may be given with advantage. This quantity may be repeated every hour or two with perfect safety, and, indeed, in many cases, it will be requisite to do so in order to accomplish the end in view. Thus, in hysteric and other convulsions, cramp in the stomach, and other spasmodic affections, if this agent be relied upon alone, it will be requisite to give it in full and repeated doses.

Caulophyllin is a remedy combining a number and variety of therapeutic properties, or at least capable of producing a change of action in a variety of morbid conditions, which change results in the restoration of a physiological condition. Amenorrhea, that is, simple amenorrhea, is successfully treated with Caulophyllin. THREE to FIVE grains three times per day will meet the necessities of most cases. When complications exist, suitable combinations may be formed with other of the Concentrated Medicines. With this, as with many other remedies, we have found alternation the most successful plan of treatment. Thus, if there be hepatic aberation, we give one of the following powders twice or thrice a week:

℞.
    Podophyllin ........................grs. Vj.
    Asclepin...........................grs. Xij

Mix and divide into six powders. These we direct to be taken at night, and the Caulophyllin three times daily. If the case is obstinate, or has become chronic before application is made for treatment, we vary the prescription. We then combine the Caulophyllin with Senecin, as follows:

℞.
    Caulophyllin,
    Senecin...........................aa. ℈ij.

Mix and divide into twenty powders. Dose, one, three times daily. We also modify the combination of the Podophyllin, thus—

℞.
    Podophyllin,
    Baptisin,
    Asclepin .......................aa. grs. X.

Mix, and divide into ten powders. One to be exhibited every second or third night, same as above. If much nervous derangement be present, the addition of from ONE FOURTH to

ONE HALF grain of Gelsemin to each dose of the Caulophyllin and Senecin will answer an admirable purpose.

In amenorrhea occurring in anemic habits, we know of no better general remedy than the following. We have used it in a large number of cases with complete success.

℞.
    Caulophyllin -     ·     ·     ·     ·
    Senecin     ·     ·     ·     ·     ·     aa. ℈ij.
    Iron by Hydrogen   ·     ·     ·     ·   grs. X.

Mix, and divide into twenty powders. Give ONE, three times per day. In many cases the quantity of Iron may be increased to ONE grain three times a day with decided advantage. In some cases the Phosphate of Iron may be substituted for the Iron by Hydrogen, and may, perhaps, answer a better purpose. This will be the case when there is much tendency to wasting of the tissues, provided no gastric irritation be present. If hysteric symptoms be present, the Valerianate of Iron may be used with advantage. In dysmenorrhea, the Caulophyllin is an admirable remedy, both for the relief of the present symptoms, and for the radical alleviation of the derangement. It is a special alterative and tonic to the uterine system, regulating and giving tone to the functions of that organ. It relieves the distress attendant upon dysmenorrhea, and its continued use during the inter-menstrual period will prove a prophylactic in a large majority of cases. When Caulophyllin is not sufficient of itself to give relief, we combine it with Viburnin and Gelsemin, as follows:

℞.
    Caulophyllin -     ·     ·     ·     ·
    Viburnin   ·     ·     ·     ·     aa. grs. XX.
    Gelsemin   ·     ·     ·     ·     grs. V.

Mix, and divide into ten powders. Dose, ONE, every two hours until relieved, or, in severe cases, every hour. This we deem as near a specific as any medicine can be, in these cases.

For menorrhagia, we have found the Caulophyllin one of the most effective of the vegetable agents. It should be given in suitable doses during the intermenstrual period, and when the menses are present in connection with Oil of Erigeron, Trillin, Lycopin, etc. It may be combined with Helonin, as follows:

℞.
 Caulophyllin .......................... ℈ij
 Helonin .............................. ℈j.

Mix, and divide into twenty powders. Give one three times daily. This will be found an excellent combination. Also with Senecin, as directed for amenorrhea.

In the radical treatment of hysteria, Caulophyllin will be found a valuable auxiliary. It may be given alone, or in combination with Cypripedin, Scutellarin, Lupulin, Hyoscyamin, or Gelsemin, etc. Combined with one or more of these, and alternated with tonics, as Cerasein, Cornin, Hydrastin, Populin, or Iron, the most desirable results may be anticipated.

For the relief of after-pains, the Caulophyllin will be found efficient in a large number of cases. If not, the combination recommended for dysmenorrhea will seldom fail. Other combinations may be effected with suitable agents, at the option of the practitioner.

The Caulophyllin has gained considerable repute as a parturifacient, and our experience in its use has fully confirmed our previous estimate of its utility. For quieting and harmonising the action of the uterus, and of the nervous system generally, relieving cramps, and other unpleasant symptoms, it is a perfectly safe, and a generally successful, remedy. It is employed by many as a partus accelerator, and, by some, preferred to Macrotin. Many practitioners are of opinion that it acts more promptly upon the uterine system than the Macrotin.

As an auxiliary in the treatment of leucorrhea and gonorrhea, it is deservedly held in high esteem. It is seldom relied upon alone, but usually employed as an adjunctive to other remedies.

Caulophyllin is an admirable remedy in some forms of dyspepsia, particularly those cases attended with spasmodic symptoms. Where there is gastric irritability, and vomiting of the food, the Caulophyllin may be employed with advantage when more decided tonics would aggravate the symptoms. If laxatives are needed, it may be employed in connection with Leptandrin, or Juglandin, or Euonymin.

Caulophyllin is employed in connection with other remedies, in the treatment of rheumatism, both acute and chronic, with much benefit. It is mildly diaphoretic and diuretic, hence appropriate in that disease as an alterative and promoter of depuration. It is frequently useful in allaying the spasmodic pains accompanying that complaint. When combinations are desired, it may be used in connection with Asclepin, Gelsemin, Veratrin, Hyoscyamin, etc., in the acute form, and with Macrotin, Sanguinarin, Xanthoxylin, Phytolacin, etc., in chronic cases.

In dropsy, it is mainly useful as a general alterative, gently stimulating absorption, diaphoresis, and diuresis. It also proves a tonic to the digestive apparatus, and so becomes instrumental in restoring the tone of the system.

Caulophyllin has been employed with much benefit in hooping cough, asthma, and for the relief of hiccough. In hooping cough, it operates well in combination with Asclepin. In asthma, with Macrotin, Gelsemin, Apocynin, etc. In apthous sore mouth, both as a gargle and as an internal remedy, the Caulophyllin has been highly spoken of. It may be used in connection with Myricin, Baptisin, Rhusin, and other appropriate remedies.

The Caulophyllin has gained considerable repute as a vermifuge, but upon this point we are not prepared to speak positively. Certain it is, that during its exhibition for other disorders, worms have been expelled in considerable numbers, giving good grounds for supposing it instrumental in their expulsion. It is deserving of further trial in this respect. If auxiliary agents are desired, Chelonin, Apocynin, Santonin, Gelsemin, etc., may be employed, according to the variety of entozoa suspected of being present. As the Caulophyllin is slightly astringent, it will be necessary to administer a cathartic occasionally during the use of that remedy. In all cases of debility, spasms and convulsions arising from uterine derangement, nervous irritability, chorea, etc., occasion will be had for the employment of the Caulophyllin, and much confidence may be reposed in its remedial value.

To sum up the history of the Caulophyllin, we would recommend it as being useful, in addition to the complaints above enumerated, in passive hemorrhage, congestive dysmenorrhea, epilepsy, nervous headache, neuralgia, hypochondriasis, prolapsus uteri, and as a general alterative remedy in all vitiated conditions of the system. Also as an agent for modifying the action of Podophyllin, preventing griping, expelling flatulence, etc. A *narcotic* property is attributed to the Caulophyllin by some writers, but we have never been able to discover it, although we have prescribed this remedy extensively during the past five years. We are of opinion that the statement was put forth by some one having a *theoretical* acquaintance only with the therapeutic history of the Caulophyllin.

# JALAPIN.

Derived from *Ipomœa Jalapa.*
Nat. Ord.—*Convolvulaceæ.*
Sex. Syst.—*Pentandria Monogynia.*
Common Name.—*Jalap.*
Part Used.—*The Root.*
No. of Principles—*one,* viz., a *resin.*
Properties.—An *irritant hydrogogue cathartic.*
Employment.—*Dropsy, fevers,* and whenever a powerful *local cathartic* is indicated.

THE medical properties and uses of both the Jalap root and its active cathartic constituent, Jalapin, are so well and generally understood, that but little is left for us to say. The Jalapin, as will be observed, consists of a single resin principle, which embodies the cathartic power of the plant. The plant, however, yields another principle, a neutral, first obtained by Messrs. B. Keith & Co., in the form of a beautiful cream-colored powder. This principle is perfectly soluble in water, devoid of cathartic properties, and powerfully diuretic. It may be inquired why a deviation is made in favor of this remedy, in not combining the two principles in the Jalapin offered to the profession. The reason is simply this—practitioners of medicine are **not** so over-stocked with wisdom as not to be sometimes deluded by outside appearances, in which respect they are so much like the rest of mankind that we can see no difference. The jalap *resin* is *white*, while the *neutral* principle is of a dirty cream color, and mixing the two together would not improve the appearance of the neutral, while it would completely destroy the immaculacy of the resin. Now the profession have hitherto been supplied with the Jalap resin, and have never known anything of the existence of a neutral principle, hence the difference of shade became a stumbling-block to honest practitioners, and a sweet nut for malicious scribblers, out of which to crack the charge of fraud and adulteration. Consequently the resin alone was put up as the *equivalent only* of the *resin* of Jalap already before the profession. We have no doubt but what the time will soon come when the combined principles of the Jalap will be as eagerly sought after as those of other plants. Certainly, if it be desirable to have a *concentrated equivalent* of the plant, such must be the case. The active diuretic properties of the neutral principle, combined with a very mild laxative power, renders it desirable in dropsy, in which disease the Jalapin is so frequently employed.

The Jalapin is employed in all cases in which it is desirable to produce a speedy evacuation of the bowels. It is contraindicated in all cases accompanied with gastric or enteric inflammation. It usually produces much tormina during its operation, which may be prevented in a measure by combining it with stimulants and anti-spasmodics, as Capsicum, Ginger, Xanthoxylin, Caulophyllin, etc. Where Podophyllin or other cathartics do not operate promptly, as is frequently the case in cold, asthenic forms of disease, we exhibit the Jalapin in doses of from TWO to SIX grains, for the purpose of relieving intestinal engorgement. It may be combined with capsicum or ginger, or what is better, a tea of ginger may be taken freely during its operation.

The average dose of the Jalapin is THREE grains. It is very seldom used alone, except in the cases above mentioned. As stated under the head Podophyllin, it is frequently combined with that remedy when it is desirable to produce a speedy evacuation of the alimentary canal. The Jalapin will generally operate in two hours, while the Podophyllin will take its own time, being, so far as we have been able to discover, neither quickened nor in any other way influenced in its action by the Jalapin.

The most powerful hydrogogue cathartic we have ever employed in dropsy, is the following:

℞.
    Jalapin,
    Podophyllin........................aa. grs. ij.
    Cream of Tartar .........................℈j.

Mix. Give at a dose. The quantity of Cream of Tartar may be increased to ONE drachm if thought desirable. This combination is admirably calculated to arouse the action of the liver, and to powerfully stimulate the entire glandular system. It is of particular service in cases of dropsical effusions into the larger cavities, as ascites, hydrothorax, etc. In most cases of dropsical effusion, and particularly in anasarca or general dropsy, the Ampelopsin should be given in suitable doses twice or thrice a day, and the above compound powder of Jalapin administered once or twice a week. Jalapin is also much employed in hydrocephalus, hydrothorax, and cardiac dropsy, in connection with Digitalin.

In large doses, Jalapin sometimes proves emetic. The free use of mucilages and demulcents is advisable when Jalapin is administered.

Jalapin is also employed in bilious fever, congestion of the portal circle, and as a revulsive remedy in many forms of disease. Yet we have other remedies of equal efficacy in those complaints, in fact preferable, and calculated, when fully known, to supercede it.

# PHYTOLACIN.

Derived from *Phytolacca Decandria.*
Nat. Ord.—*Phytolaccaceæ.*
Sex. Syst.—*Decandria Decagynia.*
Common Names.—*Poke Root, Garget, Scoke, Pigeon Berry, Coakum,* etc.
Part Used.—The *Root.*
No. of Principles—*two,* viz., *resinoid, and neutral.*
Properties.—*Alterative, resolvent, deobstruent, detergent, anti-syphilitic, anti-scorbutic, anti-herpetic, diuretic, laxative,* slightly *narcotic,* and, in larger doses, *emetic* and *cathartic.*
Employment.—*Rheumatism, scrofula, syphilis, gonorrhea, salt rheum, itch,* and other *cutaneous diseases, glandular affections,* as *tuberculosus of the liver, spleen,* etc., *carcinoma, hepatic torpor, etc.*

IN Phytolacin, we have one of the most decided and efficient alteratives embraced in the range of the materia medica. It is not a remedy of doubtful powers, but uniform, certain, and reliable in its action. In all conditions of chronic disease, wherein there is tardiness of action on the part of the exhaling, absorbing, secreting, or eliminating vessels, or a viscid and plastic condition of the blood and fluids, the Phytolacin will be found the most efficient, as well as the safest remedy that can be brought to bear. In cold and languid conditions of the system, it will rouse an action when other remedies fail of their accustomed effects. When Podophyllin seems tardy in awakening the liver from its torpor, from ONE to TWO grains of Phytolacin, added to each dose, will be found a most desirable and efficient adjunctive. It becomes almost indispensable in the treatment of long standing disorders of the liver, when once its full value is known.

In doses of from ONE to TWO grains, twice or thrice a day, the Phytolacin proves a certain, safe, and effectual resolvent and alterative, manifesting its influence throughout the entire glandular system. Many systems are so sensitive as not to be able to bear more than ONE-FOURTH or ONE-HALF of ONE grain, while in other cases from THREE to FIVE grains will be required. In large doses, say from FIVE grains upwards, the Phytolacin generally proves emetic and cathartic, although it is not a desirable remedy for either purpose. Its cathartic operation is accompanied with much nausea, pain, and subsequent prostration. When employed as an alterative, if the patient be kept too long or too freely under its influence, a considerable degree of relaxation will attend its operation, and the patient will complain of prostration and debility. Hence it is desirable, under such circumstances, to combine it with stimulants or tonics, as the Xanthoxylin, Oil of Capsicum, Cornin, Cerasein, Fraserin, etc. The average dose of the Phytolacin is TWO grains.

Rheumatism is a disease affording a fair field for the employment of the Phytolacin. It is of more utility in the chronic than in the acute form. In the latter form, however, it may be employed with advantage when the febrile stage is passed, and as a prophylactic against a recurrence. In articular and mercurial rheumatism, we deem it superior to, and much safer than Iodide of Potassa. We have used it with much success in these cases, particularly in combination with

Stillingin and Xanthoxylin. We usually combine them in the following manner:

℞.
 Phytolacin,
 Stillingin,
 Xanthoxylin .............................aa. ℨj

Mix, and divide into twenty powders. Dose—one, three times daily. We sometimes vary the formula, substituting Macrotin for the Xanthoxylin, as follows:

℞.
 Phytolacin .............................. ℨj
 Stillingin .............................. ℨij
 Macrotin............................... grs. x

Mix, and divide into twenty powders. Administer same as above. Twice a week give the following powder:

℞.
 Podophyllin........................... gr. j.
 Leptandrin ........................... grs. ij.
 Gelsemin .............................. gr. ss.

Mix, and let it be taken at bed-time. The bowels should be kept in a perfectly soluble condition during the course of the treatment.

Phytolacin, in connection with tonics, is of admirable utility in the cure of scrofula. It should be given in small doses, and alternated with Hydrastin, or Cornin, or Cerasein, or Iron. If it be desirable to employ other alteratives, it will answer a better purpose to alternate them than to combine them. Among the latter we may mention Stillingin, Alnuin, Chimaphilin, Rumin, and Corydalin.

For the cure of syphilis and mercurio-syphilitic disorders, the Phytolacin is quite equal to any other organic remedy. If the patient be brought properly under its influence, and proper observance be paid to diet, regimen, and auxiliary treatment, a cure is almost certain. Care must be taken, however, that the patient's system does not become too much relaxed, which may be avoided by the use of suitable stimulants and tonics. By employing the Phytolacin for three or four days at a time, and then alternating with Corydalin for an equal period, which is of itself a decided tonic, the necessity for employing other tonics will be lessened. Smilacin will be a valuable adjunctive to the Phytolacin in the treatment of syphilis. Also Stillingin, Myricin, Irisin and Ampelopsin. One or more of these agents may be combined with the Phytolacin, at the discretion of the practitioner. The severe pains attending tertiary syphilis, and mercurio-syphilitic complications, are more effectually relieved by the use of the Phytolacin than by any other remedy. In these cases it may sometimes be advantageously employed in connection with Hyosciamin.

Gonorrhea and leucorrhea have been successfully treated with Phytolacin. It is peculiarly serviceable in cases of long standing.

Salt rheum, itch, and other cutaneous eruptions, have been cured with Phytolacin. It is employed not only internally but externally. It may be made into an ointment or tincture.— FIFTEEN GRAINS of the Phytolacin may be rubbed up with one ounce of lard, or dissolved in one ounce of alcohol, which may be diluted with water before applied. Both the ointment and tincture have been found useful in piles. The Phytolacin possesses considerable discutient power, and the ointment applied to tumors, glandular swellings, etc., will frequently discuss them.

Phytolacin has been found of service in tuberculous affections of the lungs, liver, spleen, mesentery, etc. In the absence of febrile excitement, it is always appropriate in glandular diseases of whatever type. Its efficient alterative and resolvent properties render it valuable in promoting the absorption of all abnormal exudations and deposits.

Phytolacin has been much employed in the treatment of carcinomatous affections. It is, undoubtedly, as efficient an alterative as can be safely employed in that disease. Its beneficial effects are most apparent in cases of open cancer. The patient's system should be brought fully under its constitutional influence, and the dry Phytolacin applied to the ulcer. It may be used either alone or combined with Hydrastin, equal parts. To relieve the fœtor of cancerous sores, it should be combined with Baptisin. The Phytolacin, applied either in the form of a paste with water, or in strong alcoholic tincture, has been found quite effectual in that species of cancer known as *lupus*, when used in the early stages. Also in removing warts and corns. The strength of the ointment and tincture above directed for external application may be varied to suit occasion, being careful not to apply it too freely when an extensive abrasion of the surface exists.

# HYOSCYAMIN.

Derived from *Hyoscyamus Niger*.
Nat. Ord.—*Solanaceæ*.
Sex. Syst.—*Pentandria Monogynia.*
Common Name.—*Henbane*
Part Used.—*The Herb.*

No. of Principles—*four*, viz., *resin*, *resinoid, alkaloid,* and *neutral.*

Properties.—*Anodyne, antispasmodic, soporific, sedative, narcotic, diuretic,* and *laxative.*

Employment.—*Neuralgia, gout, rheumatism, asthma, hooping cough, croup, chronic cough, hyperæsthesis, cramps, convulsions, nervous pains, catarrhal affections, bronchitis, laryngitis, etc., etc.*

PERHAPS no other agent of the materia medica is better calculated to illustrate the defects of so-called officinal preparations than the Hyoscyamus Niger. The various pharmaceutical preparations of this plant, such as tinctures, extracts, etc., are in the highest degree uncertain and unsafe, as we shall endeavor to demonstrate. The same objections pertain to this as to all other crude medicines. In the first place, the actual amount of active constituents residing in the plant is variable, indefinite, and uncertain. In the second place, these constituents are very susceptible to disintegrating influences, and readily undergo the destructive decomposition described in the first part of this work. The extracts of this plant generally become inert and worthless within six months after they are manufactured. In the third place, the total therapeutic value of the plant does not reside in *one*, but in *four* distinct proximate active principles, each one representing therapeutic properties peculiar to itself. These several principles are of different solubility, requiring different menstrua for their extraction, and the variation in their proportions, or the absence of one or more principles in the ordinary preparations, and which is almost universally the case, renders them not only of uncertain therapeutic value, but also unsafe. This fact will be apparent when the diverse properties and influences of the several principles are considered. Thus the alkaloid principle, the *hyoscyamine* of some writers, has but very little of that peculiar effect upon the epidermis so characteristic of the plant, while it possesses the diuretic power in a high degree, and also the narcotic, or that property which chiefly affects the brain and has a tendency to produce cerebral congestion. The resin embodies the relaxant and anti-spasmodic properties to a much fuller extent than the other principles, while the neutral is mainly diaphoretic. It will be seen, therefore, how important it is, in order to realise the true and full therapeutic character of the Hyoscyamus, that its pharmaceutical preparations should contain all the active medicinal constituents of the plant, and that they should be of definite and uniform medicinal strength.

In medicinal doses, Hyoscyamin acts as a powerful sedative to the nervous system, lessens impressibility to irritation, and obviates those conditions of morbidly exalted sensibility so frequently observable in disease, while, at the same time, it increases the activity of the secreting apparatus, particularly of the glands, mucous membranes, skin, kidneys, and bowels.

In larger doses it produces dryness of the mouth and throat, thirst, nausea, vertigo, deafness, and headache. At other times, a dull, heavy feeling in the head, debility, confusion of the ideas, optical illusions, dilatation of the pupils, with increased heat of the head, and coldness of the extremities. The extremities, and particularly the tongue, become partially paralysed and immovable. These symptoms are often accompanied with great difficulty of breathing, anxiety, etc.

In very large doses Hyoscyamin produces severe convulsions, tetanic cramps, swooning, coma, paralysis, and apoplexy. When given to persons of a full, plethoric habit, Hyoscyamin stimulates the arterial system, but in general reduces the force and frequency of the pulse. The secondary effects of large but not fatal doses of Hyoscyamin are manifested by increased and copious perspiration and expectoration, and frequently a slight ptyalism. The autopsy in those cases in which Hyoscyamin has proved fatal seldom reveals any real inflammation of the stomach. The veins and blood vessels of the head are generally injected with much dark blood, and also the lungs. The blood exhibits the appearance of undergoing decomposition, and the cadaver rapidly putrefies.

Hyoscyamin acts most promptly and energetically when brought in direct contact with the cell-substance, or injected into the veins. When injected into the rectum, sudden, violent, and serious results have been witnessed.

Hyoscyamin is considered anodyne and anti-spasmodic. It depresses the sensibilities of the nervous system, and lessens the irritability of the fibres. Although affecting the brain to a greater or lesser extent, it seems, by preference, to expend its influence chiefly upon the peripheral nervous system, upon the nervous structure of the epidermis, and upon the nerves of sensation. It promotes the action of the cutaneous exhalents, of the lungs and mucous membrances generally, and also of the glandular structure, kidneys, etc. In view of its influences in these respects it is accredited with resolvent powers.

Hyoscyamin is generally employed in hyperæsthesis, nervous

pains and spasms, erethismus, and febrile conditions of the vascular system, particularly when arising from increased irritability of the nerves of sensation. In catarrhal, and even in inflammatory affections of the mucous membrances of the respiratory organs, it is used with much success.

Hyoscyamin is of service in the treatment of nervous fevers of an erethismal character, but is contra-indicated in cases of vital or paralytic debility. It is valuable for the relief of hyperæsthesia, morbid acuteness of the organs of sense, phantasma, and their accompaniments, nervous irritability and wakefulness. Also in the treatment of local inflammations complicated with idiopathic or secondary symptoms of exalted nervous sensibility, manifested by pains of an unusually violent character, with much spasmodic action, as, for instance, nervous and catarrhal inflammation of the lungs, bronchitis, laryngitis, pharyngitis, etc. As an adjunctive remedy in the treatment of croup, it has been of much value, as well as in obstinate catarrhal coughs, and in the early stages of hooping cough. In hemoptysis, when anti-spasmodics are indicated, preference is given by many to the Hyoscyamin. For the same reason it is appropriate in other hemorrhages accompanied with spasmodic action.

In consumption of the lungs Hyoscyamin is frequently of essential service, moderating the spasmodic and erethismal symptoms, and gently promoting expectoration.

Amongst the nervous affections in which the Hyoscyamin has been used with much success are included all those cases accompanied with hyperæsthesis. In mania and melancholy, when there is an abnormally exalted condition of the sensibilities, painful acuteness of touch and other senses, phantasma, and kindred symptoms, as well as in natural somnambulism, precocious development of the sexual functions, nymphomania, etc., the Hyoscyamin will be found an invaluable auxiliary remedy. In these cases it is the surest and safest of the narcotic remedies. Hyoscyamin is also of great service in the treatment of amaurosis arising from excessive nervous sensibility, nervous headache, facial neuralgia, and nervous toothache. In general convulsions, accompanied with hyperæsthesis, arising from an erethismal condition of the nervous system, and unaccompanied with fever or cerebral excitement, and in epilepsy, hysteric tetanus and trismus, chorea, etc., Hyoscyamin is employed with much success. Also in the convulsions of nursing children, particularly those arising during dentition.

Hyoscyamin is contra-indicated in acute sanguineous inflammations, vital or paralytic debility, violent determinations to the head, dyscrasia, and in all diseases having a putrefactive tendency.

Externally, the Hyoscyamin is sometimes employed as a local application in various inflammatory, spasmodic, and painful affections, as, for instance, painful and irritable ulcers, enlarged glands, inflammation of the mammæ, etc., in which by virtue of its relaxant, anti-spasmodic, and anodyne properties, it is frequently of much service. The dry powder may be sprinkled upon the surface of open ulcers, being careful not to use it too profusely, or applied by means of a poultice. In the latter form it is sometimes applied to the abdomen in cases of colic, and to other parts for the relief of spasmodic pains. For the purpose of applying to painful tumors and enlarged glands, it may be made into an ointment with lard. Its injection into the rectum is considered a dangerous experiment.

The dose of the Hyoscyamin will vary from ONE EIGHTH to ONE grain. It is always well to commence its use in small doses, and increase if occasion requires. To ensure a prompt and harmonious action, it should be rendered as diffusible as possible, which may be accomplished by trituration, or by the free use of diluents. We mention no combinations because we consider it a remedy of peculiar and sufficient potency in itself, and believe that the indications for its employment will be better subserved by employing the remedy uncombined, alternating with such other medicines as the necessities of the case demand. Neutralise undue acidity of the stomach previous to its exhibition. Asclepin will be found an excellent article with which to triturate the Hyoscyamin.

## CONCENTRATED TINCTURE HYOSCYAMUS NIGER.

Like the other concentrated tinctures of which we have already spoken, this preparation represents the entire therapeutic value of the plant in a condensed and reliable form, and of definite and uniform medicinal strength. It is very convenient for office dispensation, and for combining with other of the concentrated tinctures. We employ it more frequently than the Hyoscyamin. In fevers and other acute diseases, when not contra-indicated, we find it of great value in relieving pain and spasm, and procuring sleep. In acute rheumatism, and in scarlatina, measles, pneumonia, etc., we have derived much satisfaction from its employment. In menorrhagia, dysmenorrhea, and similar affections, it will be found a valuable anti-spasmodic and anodyne. For the relief of those peculiar headaches arising from an anæmic condition of the system, we know of nothing to equal it. Also for allaying excessive irritability of the nervous system arising from excessive hemorrhages, or profuse colliquitive discharges. Useful combinations may be effected with the Con. Tinc. Senecio, or Scutellaria, or Gelseminum, or Veratrum, etc., when desired.

Locally, it may be applied by means of lint, or otherwise, to painful tumors, enlarged glands, and in cases of local neuralgic pains, rheumatic swellings, cramps, colicky pains in the abdomen, etc.

The dose of the concentrated tincture will vary from FOUR to TWENTY drops, and even more. It may be repeated, in severe cases, once in two hours.

# STILLINGIN.

Derived from *Stillingia Sylvatica*.

Nat. Ord.—*Euphorbiaceæ*.

Sex. Syst.—*Monœcia Monodelphia*.

Common Names.—*Queen's Root, Queen's Delight, Yaw Root, Marcory, Cock-up-hat,* etc.

Part Used.—*The Root*.

No. of Principles.—*Four*, viz., *resin, resinoid, alkaloid* and *neutral*.

Properties.—*Alterative, resolvent, stimulant, tonic, diuretic, anti-syphilitic,* etc.

Employment.—*Scrofula, syphilis, leucorrhea, gonorrhea, cutaneous diseases, incontinence of urine, impotence, sterility, rheumatism, bronchitis, stomatitis,* and whenever an *alterative* is required.

THE Stillingia Sylvatica has long been in use in popular practice, but it is only of late that its remedial value has been duly recognised by the profession. In addition to the proximate active principles above enumerated, the plant also yields an oil, which will be treated of in the proper place.

We believe that the Stillingin now offered to the profession by Messrs. KEITH & Co., embodies the therapeutic value of the plant in the most condensed and reliable form of any hitherto prepared. This opinion is based upon an observation of its utility in the treatment of disease. As an alterative, it has few, if any, superiors. The average dose of the Stillingin s THREE grains. When used alone, this quantity may be repeated three times a day. The dose must be varied to meet the peculiarities of the case, as some patients will require double, and even quadruple the quantity of others to produce the desired effect. In over-doses, it will produce nausea and sometimes vomiting. The proper time to administer it is two hours after meals. If taken a short time before meals, it materially interferes with the appetite.

Among the diseases in which the Stillingin has been found most efficient, we might mention scrofula, gonorrhea, syphilis, leucorrhea, rheumatism, and mercurial affections. In order to realise its full utility, when used alone, its use must be persevered in for a length of time. From TWO to FOUR grains may be given three times a day in scrofula, the bowels being kept in a soluble condition by small doses of Podophyllin, or Leptandrin, or Euonymin, etc. It is well to alternate the Stillingin occasionally with other alteratives. Or, if preferred, suitable combinations may be effected with other remedies.

For gonorrhea and syphilis, the Stillingin is usually employed in combination with other agents, as the Corydalin, Irisin, Phytolacin, Smilacin, Myricin, etc. It is better, in these cases, to premise the alterative course with a thorough dose of Podophyllin, which will prepare the system for the action of alterative remedies, and which should be repeated at suitable intervals during the treatment. One fact we have observed, in connection with the employment of Stillingin in the treatment of gonorrhea, and that is, its tendency to provoke urethral irritation and chordee, rendering its use, in some cases, inadmissible. This property, however, renders it of great value in the treatment of incontinence of urine, impotence, and sterility. In all atonic and paralytic affections of the generative and urinary apparatus, it seems to be a remedy of much value. The most obstinate cases of leucorrhea have yielded to the Stillingin.

Chronic rheumatism affords a fair field for the successful employment of this remedy. It may be used alone, or combined with such other of the concentrated agents as are suited to the case. In several cases of articular and mercurial rheumatism, we have used the following formula with much benefit

℞.
    Stillingin............................grs XX
    Irisin................................grs. X.
    Phytolacin..........................grs. V.

Mix, and divide into ten powders. Give one three times per day. Or the following:

℞.
    Stillingin,
    Xanthoxylin....................aa. grs. XX.
    Macrotin..........................grs. V.

Mix and divide into ten powders. Doses same as above. Both these formulas will be found excellent. When a mild laxative is indicated, the Menispermin will answer a good purpose. They may be combined in equal proportions. If a more energetic remedy of this class is called for, Euonymin will be found admirably suited to the occasion. Few remedies excel the latter when a laxative tonic is required.

We would not be understood to say that the Stillingin is fully equivalent to the plant, as considerable of its medicinal value resides in the oil, of which we next propose to treat. Deprived of the oil, Stillingin is not so efficacious in the treatment of affections of the respiratory organs, nor of leucorrhea and other kindred female diseases. Nevertheless, it is a valuable stimulating alterative, exciting the glandular system in a peculiar manner, resolving viscidity of the secretions, and promoting depuration. It is of great utility, in combination with Xanthoxylin, in the convalescing stages of cholera infantum, dysentery, and other diseases attended with colliquitive discharges. They should be combined in equal proportions, and administered in four grain doses three or four times a day.

For paralytic affections of the bladder, it may be employed with much confidence. In this affection, it may be used in conjunction with electricity, with much prospect of benefit.

Chronic diarrhea and dysentery have been cured with alternate doses of Stillingin and Leptandrin. From two to four grains of Stillingin may be given twice or thrice daily, and the same quantity of Leptandrin at bed time. In cold and sluggish conditions of the system, Stillingin operates well in combination with Macrotin. In chronic diseases of the liver, with Euonymin, Phytolacin, etc. In the treatment of dermoid diseases, Stillingin is justly esteemed a remedy of great value. Average dose, THREE grains.

# OIL OF STILLINGIA.

Derived from the root of Stillingia Sylvatica.

We deem this remedy one of the most valuable accessions to our indigenous materia medica. Although pronounced by some authors to be too acrid for internal use, we have found such not to be the case. We have employed it largely in bronchitis, laryngitis, and other affections of the respiratory system, and in defective menstruation, chronic gleet, leucorrhea, etc., and have found it a remedy of safe and exceeding utility.

The average dose of the oil is ONE drop, which may be repeated every half hour, in croup, with safety. In other cases, every four or six hours. It may be dissolved in alcohol, and taken in a little water, or dropped upon sugar, or mixed with mucilage of gum arabic, slippery elm, etc. We are of opinion that its local action is most beneficial in bronchitis and laryngitis, hence prefer to administer the oil upon a little sugar, which may be allowed to dissolve in the mouth and gradually swallowed. The following will be found an elegant and efficient remedy for coughs, colds, bronchitis, influenza, etc.

℞.
    Oil Stillingia .......................... ʒj.
      " Wintergreen
      " Cinnamon........................aa. gtt. X.
    Hydrastin........................... Ɔj.
    Alcohol............................. ʒ X.

Mix Dose—from TEN to FIFTEEN drops four or five times a day, or whenever the cough is troublesome. The addition of ℥ ss. of Oil of Xanthoxylum will improve the mixture for cases of long standing.

We have administered the Oil of Stillingia in croup with marked advantage, our first experience having been in our own family. It seems to operate as a powerful diffusible stimulant, resolvent, and anti-spasmodic. It overcomes the spasm and difficulty of respiration, and favors expectoration, hence will be found useful in asthma, hooping cough, and other kindred affections. For the relief of asthma, it may be combined with Oil of Lobelia.

We have employed the Oil, in combination with Oil of Erigeron, with the most gratifying success in the treatment of defective menstruation. The Oils may be combined in equal proportions, and from ONE to THREE drops taken three times a day. The peculiar headaches accompanying this affection are soon relieved by the use of this remedy. The same combination will be found of great utility in uterine leucorrhea, and in gonorrhea. When it is desirable to have the entire properties of the Stillingin combined, the following formula must be observed:

℞.
    Stillingin............................... ℥ ij.
    Oil of Stillingia........................ ℥ ss.
    Alcohol, 95 per cent..................... ℥ X.

Mix. Dose—from TEN to FIFTEEN drops. This secures the entire therapeutic value of the plant, and constitutes one of the most efficient remedies known for the cure of scrofula, syphilis, eruptions of the skin, and all affections arising from a vitiated condition of the blood.

Externally, the oil is an invaluable stimulant, counter-irritant, and relaxant. It relaxes spasm of the muscular fibres and at the same time stimulates the depurative functions of the skin to healthful activity. Among the affections in which it may be employed with certainty of benefit, we might mention croup, asthma, acute and chronic pleuritis, pneumonia.

neuralgia, spinal affections, contracted joints, etc. For external use, it should be dissolved in alcohol, the proportions varying according to the degree of stimulation required. In ordinary cases, we observe the following proportions:

℞.
    Oil of Stillingia............................ ʒj.
    Alcohol 95 per cent........................ ℥j.

Mix. Bathe the affected parts freely two or three times a day, or apply a cloth saturated with the solution. For slight neuralgic affections, spinal irritation, and rheumatic pains, this will be found of great service. We employ it, however, most frequently in combination with the Oils of Lobelia and Capsicum. Our formula is as follows:

℞.
    Oil of Stillingia ......................... ʒj.
    "    Lobelia............................. ʒss.
    "    Capsicum ..........................gtt. XX.
    Alcohol 95 per cent....................... ℥ij.

Mix. This we esteem one of the most valuable external applications ever devised. The quantity of the oils may be doubled, or even trebled, to meet the indications in very severe cases. In case more of the counter-irritant property is desired, the quantity of Stillingia may be increased, and the other ingredients allowed to remain the same. For croup, hooping cough, and asthma, bathe the throat and upper portion of the chest with this preparation two or three times a day. Its employment will be followed, after a few days, by a profuse vesicular eruption, which, in a few days, will assume a pustular character. Frequently the eruption will appear within six hours after the first application. Spinal irritation, neuralgia, tic doloreux, rheumatic pains, contracted joints, chronic sprains, etc., are relieved and cured by the use of this remedy. When the relaxant property is not needed, the Oil of Lobelia may be dispensed with. We sometimes vary the formula, thus:

℞.
 Oil of Stillingia
 "  Erigeron........................... ℨj.
 "  Lobelia............................ ℨss.
 Alcohol................................. ℥ij.

Mix. This formula is peculiarly serviceable in bronchitis, laryngitis, enlargement of the tonsils, rheumatic pains, etc. The Oil of Cajeput may be substituted for the Erigeron when the latter is not at hand. It will be seen that the combinations may be easily varied, according as more or less stimulating or relaxing applications are required. We are certain that those who once test the value of the Oil of Stillingia as an external remedy, will be loth to dispense with it.

# LUPULIN.

**Derived** from *Humulus Lupulus.*
Nat. Ord.—*Urticaceæ.*
Sex. Syst.—*Diœcia Pentandria.*
Common Name.—*Hops.*
Part Used.—*The Strobiles, or Cones.*
No. of Principles—*three,* viz., *resin, resinoid, and neutral.*
Properties.—*Nervine, hypnotic, febrifuge, diuretic and tonic.*
Employment.—*Dyspepsia, delirium tremens, hysteria, after-pains, chordee, spermatorrhea, intermittent fevers, etc.*

The Lupulin under consideration should not be confounded with that usually found in commerce, which consists simply of the pollen of the flowers. In the Lupulin of which we propose to treat, we have not only the virtues of the pollen, but also additional properties derived from the parenchyma of the flowers. Lupulin is a remedy of much value in the treatment of nervous affections and is frequently employed as a substitute for opium, possessing the advantage of not disturbing the stomach, or producing constipation. Like all remedies of its class, however, it is not always to be relied upon for the purpose of allaying nervous excitement, frequently failing of its influence in this respect. In such cases it proves mainly diuretic.

The average dose of the Lupulin is TWO grains, increased to FIVE with benefit. On account of its febrifuge properties, it is peculiarly appropriate in the treatment of febrile diseases for the purpose of controlling the excitability of the nervous system, and correcting a tendency to delirium. It will frequently procure refreshing sleep in cases of great wakefulness when other remedies fail. In many cases it is an invaluable anodyne, allaying pain, promoting diaphoresis and diuresis, and inducing sleep. It has been used with good results in delirium tremens. In this complaint larger doses than usual are required, TEN grains, repeated every two hours, having been administered with success Nervous headaches, hysteria, chronic cough, suppression of urine, and various other affections have been relieved and cured by the use of Lupulin. In those forms of indigestion wherein there is a tendency to gastritis, the Lupulin will be found an excellent remedy. It soothes and allays the irritability of the mucous tissues, and paves the way for the employment of more decided tonics. In these cases it is beneficially administered in combination with Helonin.

℞.
    Lupulin .......................... grs. XX.
    Helonin .......................... grs. X.

Mix, and divide into ten powders. Give one three times per day. Or with Smilacin.

℞
    Lupulin,
    Smilacin, ........................ aa. ℈j.

Mix, and divide into ten powders. Dose—same as above. In chronic gastritis, enteritis, and ulcerations of the stomach and bowels, the latter formula will be found useful.

Lupulin has been used with extraordinary success in the

cure of spermatorrhea. From TWO to FIVE grains are given at a dose, and repeated three or four times daily. Some practitioners use it in combination with Cerasein, and with marked advantage:

℞
    Lupulin ......................................... ℈j.
    Cerasein........................................ ʒss.

Mix and divide into ten powders. Dose—one, once in six hours.

The efficacy of the Lupulin in the treatment of spermatorrhea is enhanced by combining it with Gelsemin, and alternating with Cerasein. We prefer the following method of administration:

℞
    Lupulin.....................................grs. iij.
    Gelsemin..............................grs. ss. ad.j.

Mix. To be given at bed time, and FIVE grains of Cerasein administered three times daily. If ulceration of the urethra be suspected, use the following injection:

℞
    Chloride of Lime .......................... ʒss.
    Hydrastin................................. ʒj.
    Water .....................................O.j.

Digest and filter. This injection is valuable in gonorrnea, gleet, leucorrhea, and other affections of the mucous surfaces of the generative apparatus.

So far as our experience goes, the Lupulin here treated of may be relied upon for all the purposes for which the plant and its preparations have hitherto been employed. It has been reputed useful in the treatment of ague and fever, but we have no well authenticated evidence of its utility in that complaint. Its tonic powers are feeble at best, and seem to be expended mainly upon the stomach. In cases of suppression and retention of urine, it sometimes affords most desirable relief. Its employment is more indicated in sthenic than in asthenic conditions of the system. After-pains are frequently relieved by its use, and the nervous irritability peculiar to parturient females allayed and overcome.

Lupulin has frequently proved successful in the treatment of chordee, by virtue of overcoming the urethral inflammation, and correcting the acridity of the urine. It has the reputation of diminishing the quantity of lithic acid in the urine.

# VERATRIN.

**Derived from** *Veratrum Viride.*
Nat. Ord.—*Melanthaceæ.*
Sex. Syst.—*Polygamia Monœcia.*
Common Names.—*American Hellebore, Swamp Hellebore, Itch Weed, Indian Poke, etc.*
Part Used.—*The Root.*
No. of Principles—*four, viz., resin, resinoid, alkaloid,* and *neutral.*
Properties.—*Emetic, cathartic, diaphoretic, expectorant, nervine, antispasmodic, arterial sedative, alterative, resolvent, febrifuge, anodyne, soporific, etc.*
Employment.—*Intermittent, remittent, typhoid, and other fevers, pneumonia, pleuritis, rheumatism, delirium tremens, mania, affections of the heart, both functional and organic, congestions of the portal circle, hooping cough, asthma, hysteria, cramps, convulsions, scrofula, dropsy, epilepsy, amenorrhea, etc.*

WE fully realise our inability to do justice to the value of the article under consideration, yet we shall attempt to place before our readers what positive information we possess in regard to it. It has long been, with us, a favorite remedy, and we have learned to place much reliance upon its efficacy in many disorders afflicting the human frame.

It will be seen that we have attributed to it a considerable number of therapeutic properties, all of which we shall endeavor to substantiate, as being in accordance with our experience in its employment. No other remedy of its class, with which we are acquainted, is capable of fulfilling so many indications with safety, certainty, and uniformity of action. The indications for its employment are of frequent occurrence, and its administration affords well marked and positive evidences of its practical utility. Yet, as a necessary condition of its successful employment, a correct diagnosis is essential, and the remedy must be rightly *timed*, as well as proportioned. We do not hold it a specific in any disease, yet we claim for it the possession of positive and specific therapeutic properties available and reliable whenever the proper adaptation is had.

In order that those who are not familiar with its properties, and employment may have a better understanding of its range of utility, we will endeavor to describe its physiological influence upon the organism. Like Digitalin, its influences are diverse, and variously manifested upon the several divisions of the animal economy. Thus we call it an arterial sedative, as it reduces the force and frequency of the pulse. We cannot attempt to explain whether this influence is due to a property whereby a direct depression of the vital activity of the arterial system is produced, or whether it is the result of the correction of certain conditions which were the cause of the abnormally excited condition of the circulation. Certain it is that Veratrin is a powerful resolvent and deobstruent, resolving the plasticity of the blood, and of the secretions generally, while, at the same time, it promotes the activity of the absorbent, venous, and lymphatic vessels, and glands. It exercises a wonderful control over the capillary system, particularly the deep-seated capillaries, hence, in congestions of the remote tissues, is a remedy of great service. In small doses Veratrin stimulates the functions of the abdominal viscera, particularly of the stomach, liver, pancreas, and mesentery—promotes the secretion of the nervous fluids, and exercises a striking influence over the vascular structure of the abdomen, giving activity to the portal circulation, and promoting the sanguineous secretions, as the catamenia, hemorrhoidal flux, etc. It also quickens the activity of the renal secretion and cutaneous exhalations. Upon the nervous structure of the abdomen generally, it acts as a powerful stimulant, alterative and tonic.

In large doses Veratrin causes vomiting, diarrhea, and great depression of the arterial system, the pulse becoming very small and infrequent. The general sensibility of the system is also affected in a very disagreeable and violent manner. In very large doses, if not instantly ejected by vomiting, very violent symptoms are excited by the Veratrin. Copious and painful bilious vomitings, hemorrhagic diarrhea, metrorrhagia, tenesmus, pulse very small and infrequent, excessive prostration, subsultus tandinum, swooning, paralysis, convulsions, tetanus and death. The immediate cause of death in this instance is more to be attributed to the excessive irritation and exhausting excitement of the abdominal nervous structure, and the depression of the arterial system, than to any inflammation excited in the intestinal viscera.

In cases of febrile excitement, the first influence we have observed of the action of the Veratrin is, a softening of the pulse. Correspondingly, or immediately following, there is a slight elevation of the temperature of the skin, a gentle breathing perspiration ensues, and the skin becomes soft and flexible, while its temperature falls somewhat below the normal standard. These several phenomena being produced, the pulse becomes less frequent, full and regular. If the medicine be continued, considerable relaxation of the system is observable, and the pulse sinks to sixty, fifty, or even forty beats per minute. At this point vomiting usually occurs, and, in ordinary cases, the medicine must be omitted until the nausea subsides. In a great number of cases it is necessary to push the medicine to the production of emesis in order to bring the symptoms under control. As an emetic, the Veratrin operates generally with less of prostration than other remedies of its class. In most cases, when given in emetic doses, it operates very promptly, but is sometimes tardy, owing, we are of opinion, to acidity of the stomach. When the quantity of Veratrin given has reached an extent sufficient to produce emesis, the symptoms preceding vomiting are sometimes somewhat alarming. The patient becomes very pale, particularly about the lips and alæ of the nostrils, and complains of great faintness and dyspnea. Vomiting almost immediately ensues, and is free, copious, and without spasm. The pulse at first sinks considerably, but, as soon as vomiting has occurred, comes back to the normal standard, the temperature of the surface rises, a gentle perspiration breaks out, and the breathing becomes free and full. When used for the purpose of an emetic, the Veratrin should be thoroughly triturated with Asclepin or Eupatorin, (Perfo.,) and accompanied with a plentiful supply of fluid. In all cases the Veratrin should be thoroughly triturated with some one of the other concentrated medicines not contra-indicated in the case, of which we prefer Asclepin, as being most frequently admissable. The Veratrin is a medicine possessed of a high concentration of therapeutic power, and, in order to ensure its kindly operation, it should be rendered as diffusible as possible. Too great a concentration of therapeutic action upon a limited nervous surface will produce violent and serious symptoms, while the same amount of medicinal power diffused over a more extended space of impressible tissue will be productive of none other than kindly results. Extremes in medicine are always to be avoided. When too highly diluted or diffused, medicines become of negative value, their field of operation being too extended. On the other hand, when of too high concentration, the object in view is defeated by the overaction produced, and confusion of the vital manifestations, instead of harmony, ensues.

We have observed, as the result of the administration of Veratrin, when care has not been exercised in regard to neutralising undue acidity, and ensuring proper diffusion of the remedy, very singular contortions of the muscular system, particularly of the muscles of the face, neck, fingers and toes. The head would be drawn to one side, the mouth drawn down at one corner, and the facial muscles affected with convulsive twitchings. At the same time the fingers and toes would be cramped as in cholera. At times these contortions would take the form of tonic spasm, while at other times the action would similate a series of galvanic shocks, frequently of such violence as to precipitate the patient out of bed. During all this time the intellect of the patient remains undisturbed, and he is perfectly conscious of all that is going on. As soon as this spasmodic action has subsided, no further inconvenience is felt, the patient passing from under its influence unharmed. Several instances have come under our observation when preparations of the Veratrin have been taken through mistake. In one instance a large teaspoonful of the concentrated tincture was taken by a female patient of ours who supposed she was taking tincture of Valerian. Further than nausea and free vomiting, no ill effects were experienced. In another instance, in the practice of a brother practitioner, nearly a quart of a strong decoction of the recent root was taken within the period of a few hours. A considerable degree of sickness and prostration was produced, followed by copious vomiting and purging, but the patient soon recovered without having experienced any permanent mischief. We have never known of a single instance in which fatal consequences have ensued from the action of the preparations of the Veratrum, yet we have no doubt but that such a result might occur from the administration of very large doses of the Veratrin, as mentioned in the preceding pages, We have administered the preparations of the plant to children and adults of every age, and under almost every circumstance of chronic and acute disease, and we have come to view it as an indispensible agent in our practice. For the purpose of controlling the action of the heart and arterial system, stimulating the absorbent, venous, and lymphatic vessels and glands, it has no equal. Also as a resolvent in plastic conditions of the blood, and of the secretions generally. That it is an alterative and depurative remedy of more than usual efficiency, is evident from the thoroughly renovated and invigorated condition of the animal economy after having been fully subjected to its sanative influences. And it accomplishes its work without producing any disturbance of the cerebral functions, never exhibiting, so far as we have been able to discover, any narcotic influences whatever. In view of its general physiological control, this fact is somewhat remarkable, but which enhances its practical remedial value above that of all other remedies of its class.

Indications for the employment of the Veratrin are had whenever there is a disturbed condition of the circulation, either when the abnormal excitement involves the whole arterial system, or simply affects some of its single branches. This morbid exaltation is more frequently characterized by force and fulness, than by rapidity of the pulsations. This condition may arise from two causes. In the first place, from the presence of an undue quantity, or a too highly stimulating property of the natural excitants of the blood; and, in the second place, from an abnormally increased excitability of the heart and arterial vessels, even while the blood preserves its normal constitution. It is in the first named condition that Veratrin is more particularly indicated. We employ Veratrin as a stimulant and resolvent in obstructions and atonic conditions of the liver and portal system, and of the abdominal organs generally. Also for the purpose of promoting the depuration of retained and accumulated secretions, particularly of the sanguineous, as the catamenia, and in indolent conditions of the mucous membranes, and glandular and lymphatic systems. The peculiar stimulant and alterative properties of the Veratrin as manifested in its reactions upon the nervous tissues of the abdomen, render it a remedy of great value in the treatment of all forms of disease involving the abdominal ganglia, and in all cases of functional inactivity or obstinate torpor, as for instance, mental debility and insanity, convulsions, paralytic affections, &c.

Of the special employment of Veratrin in individual types of disease, we would note our observations as follows:

It is indicated in all forms of acute febrile disease manifesting a high plasticity of the blood, accompanied with a quick, full, and wiry pulse. This condition will be frequently met with in remittent and intermittent fevers, protracted and inveterate cases of which have been successfully treated with Veratrin. Obstinate quartan fevers, complicated with atrabilious obstructions, phlegmonoid affections of the abdominal viscera, debility and torpor of the nervous structure of the abdomen, or with feeble hemorrhoidal action, are relieved and cured by means of the Veratrin. In these affections it should be given in small doses combined with Podophyllin, and alternated with tonics, of which we prefer Cerasein. In rheumatic fevers the Veratrin is generally preferable to any other remedy, as it not only breaks up the fever, but also arrests the copious symptomatic sweats arising from excessive capillary congestion. In this case it should be combined with Asclepin and Cerasein, or they may be alternated.

In the treatment of every form of febrile exanthema, and particularly of scarlatina, the Veratrin is unequalled, as these types of disease are accompanied with great arterial excitement, a high degree of plasticity of the blood, and a strong tendency to the production of effusions and exudations, for the prevention or removal of which the Veratrin is of such remarkable utility. Were Veratrin of no further service than in the treatment of scarlatina, we should still deem it invaluable and indispensable. So far as our observations have gone, and they extend over a period of five years experimental use of the Veratrin, both in our own practice and in that of others, we have never yet seen a case treated with it that did not result in a perfect cure, unattended with effusions, exudations, or malignant sequela of any kind. In scarlatina, as we find it in this region, we premise our treatment, in the early stages, with Podophyllin, and afterwards rely upon Veratrin and Asclepin in combination. Seldom is further medication necessary, unless it be to meet special symptoms. To prepare the

Veratrin for use in scarlatina, it should be thoroughly triturated with Asclepin and made into solution with hot water. It may be then administered in such doses and with such frequency of repetition as may be necessary to control the disease. Our experience is in favor of administering it at intervals of two hours. When the inflammatory action is violent, it may be administered every hour in the commencement until the violence of the symptoms is subdued, and then repeated at intervals of two or three hours as may be necessary to maintain its proper influence. As soon as an intermission, full and complete, occurs, the Cerasein may be given in suitable doses, the Veratrin and Asclepin being continued at intervals of four or six hours until all danger of a return of the febrile symptoms is past. Veratrin seems appropriate in all stages of scarlet fever. We have known cases of the worst form, and in the latter stages, where the patient was in convulsions, and the medical attendant had abandoned all hope, in which the exhibition of this remedy has promptly arrested the disease, breaking up the convulsions and saving the patient. The absence of effusions, exudations, and other of the usual distressing sequents of scarlatina, when treated with Veratrin, we attribute to the remarkable resolvent, alterative and tonic power of this remedy, whereby the depurative action of the entire economy is promoted, and these retentions accumulations, and consequent congestions are prevented. It stimulates the functions of the absorbent, venous, and lymphatic vessels in a peculiar manner, and, by resolving the viscid and plastic condition of the blood and secretions, enables them to discharge their various functions fully and effectually. It is our firm conviction that the three remedies above enumerated, namely, Podophyllin, Veratrin, and Asclepin, will, when judiciously employed, cure a larger per centage of the cases of Scarlatina than any other plan of treatment yet devised. And when the patient is pronounced *red*, the term is no misnomer.

The virtues of Veratrin in the treatment of typhoid fever have been variously estimated, yet all agree in pronouncing it a remedy of great value. Differences of locality, atmospheric and other influences, previous habits and exposures, and many other causes tend to create a diversion in the special symptoms of typhoid fever, yet, in its general characteristics, it is the same. Derangement and torpor of the functions of the liver, portal vein, and of the secreting structure of the abdominal viscera generally, characterise the disease under all circumstances. A disposition to congestion of the glandular surfaces of the mucous membranes of the alimentary canal, is a constant accompaniment of typhoid fever. The sequent to this congestion is, an exhausting and frequently uncontrollable diarrhea, which hurries the patient to his grave, despite all means employed for its alleviation How important, then, that we possess a remedy that will early correct this functional aberration, and, by maintaining a proper degree of vital activity, obviate the danger of organic lesion. Not only is it necessary that the secreting apparatus be brought under the immediate influence of appropriate stimuli, but also that the secretions themselves shall be resolved and reduced to a degree of fluidity consistent with the ability of the apparatus to circulate them. A plastic condition of the blood is a marked characteristic of typhoid fever, and the neglect of early attention to this condition is the common cause of the fatality of this disease. Bleeding and other means of direct depletion serve to aggravate the existing obstructions by depriving the system of the fluid menstrua requisite in the work of resolution. In all febrile diseases there is danger of the solid secretions becoming in excess of the fluid, hence the free use of diluents is as indispensible a necessity as the employment of suitable medicines. Not only are they necessary for the resolving of the morbid deposits, but also for the solution and circulation of the remedy itself, whereby it may be enabled to reach the field of its operations. We have already dwelt at some length upon the necessity of the observance of this condition, in the first part of this volume, to which we respectfully direct the attention of the reader. Auxiliary remedies, in the treatment of typhoid fever, will be found in Podophyllin, Leptandrin, Euonymin, Euphorbin, Asclepin, Cerasein, Geranin, Myricin, etc., according to the indications present. In this, as in scarlet and all other fevers, the alkaline sponge bath should never be omitted.

As before mentioned, the value of Veratrin in the treatment of typhoid fever is variously estimated. While admitted by all who have employed it to be a valuable agent in controlling this disease, experience goes to prove that it is seemingly much more efficient in some localities than in others. In the section in which we reside it is not uncommon to see severe cases of typhoid fever broken up completely in from twenty-four to forty-eight hours; while in other sections we have the testimony of practitioners to the effect that, while it relieves the urgent symptoms and abates the violence of the disease, yet the fever will run its course for the accustomed length of time, although the danger is greatly lessened and recovery rendered more certain. What may be the reasons for this discrepancy of action, we have no present means of ascertaining. Whether it be owing to local influence, such as pertain to miasmatic districts—or to the want of proper preparatory or auxiliary treatment, are questions we do not feel competent to answer. The special points of congestion seem to vary in different sections. Thus, with us, the liver and brain chiefly suffer; while at the south and west, the bowels seem the most vulnerable point, diarrhea and enteritis being the most dangerous symptoms likely to arise. Whether this tendency to aggravated inflammation of the bowels depends upon the previous habits and circumstances of the patient's situation, or upon an immediate peculiarity of the disease itself, are questions which would require the closest scrutiny in order to effect a satisfactory explanation. It is the duty of every resident practitioner to study the local phenomena occurring within the circle of his observations, and to modify his treatment so as to meet existing necessities. It is unfortunate that we have no systematic concert of notation, by means of which a record of the effects of local influences in modifying the types of febrile forms of disease might be had for the benefit of the profession at large.

The employment of Veratrin in the treatment of diarrhea and dysentery affords occasion for some remarks in regard to the action of this remedy upon the bowels. We have seen it stated by some writers, that Veratrin is objectionable on account of its irritating influence upon this organ. Such has not been our experience. We have employed it much in the treatment of bowel complaints, and with the most happy results. In dysentery, after having premised our further treatment with Podophyllin and Leptandrin, when indicated, we give ONE-EIGHTH grain of Veratrin, or TWO drops of the concentrated tincture, every two hours until the febrile symptoms are subdued, and a proper action of the skin excited, alternating with Geranin, or other astringents, if needed. By referring to the preceding exposition of the physiological effects of the Veratrin, the reader cannot fail to perceive the appropriateness of this remedy in the treatment of all functional derangements of the abdominal viscera. With this remedy, as with all others, in order to ensure success, due discrimination must be exercised in regard to *time, quantity, repetition,* and *continuance.*

Veratrin is of exceeding utility in the treatment of meningitis, phrenitis, hydrocephalus, and cerebral difficulties generally. We have seen some of the most severe and desperate cases recover under its timely and persevering administration. The patient must be kept fully under its influence, until every vestige of inflammatory action has subsided. The auxiliary remedies are Podophyllin, Euphorbin, Asclepin, Scutellarin, Lobelia, etc.

Inflammatory affections of every kind, and particularly when of a hypersthenic character, afford indications for the employment of Veratrin. The peculiar influence of this remedy over the arterial system, and upon the absorbent, resolving, and lymphatic vessels of the system generally, renders it extremely valuable in this class of affections

Diseases of the mucous and serous membranes and glands are also successfully treated with Veratrin.

We have the joint experience of many practitioners in confirmation of the value of Veratrin in puerperal fever. When joined with Podophyllin, greater success has been had than by any other means or method of treatment. The most seemingly desperate cases have yielded to its sanative influences.

Veratrin ranks high as a remedial agent in pneumonia, pleuritis, croup, asthma, and other disorders of the respiratory system. It is one of the most reliable expectorants known. In all affections attended by dyspnea, Veratrin is of excellent service. We have used it in croup with entire success. In mucous and spasmodic croup it gives prompt relief. In membranous croup it is peculiarly appropriate, on account of its resolvent properties, lessening and overcoming the tendency to effusion of plastic lymph, and the formation of false membrane. In all inflammatory affections of the chest, the Veratrin is of exceeding utility. It relaxes spasm, lessens arterial excitement, equalises the circulation, resolves the viscidity of the secretions, promotes diaphoresis and expectoration, and imparts tone to the venous, absorbent, and lymphatic vessels, and glands generally. A consideration of these peculiar influences of the Veratrin will assist materially in determining its range of application.

Veratrin is of equal service in the treatment of chronic as of acute forms of disease. Its remarkable control over the heart and arterial system renders it eminently valuable in the treatment of both functional and organic disease of the heart, as palpitation, sternocardia, chronic pericarditis, enlargement, etc. We have used it in many cases of organic disease of that organ, with the most beneficial results. We deem it the safest, and, at the same time, the most efficient remedy that can be brought to bear in these disorders. Many cases of so-called organic affections of the heart have been cured by the use of the Veratrin, which, however, were nothing more than functional disturbances dependent upon visceral engorgements, suppressed secretions, metastasis of eruptions, rheumatism, etc.

Veratrin is of essential service in the treatment of atonic mucous hemorrhoids, false membranous formations in the intestinal tube, and other forms of phlegmatic disease of the abdominal cavities, particularly when dependent upon or accompanied with debility and inaction of the portal vein and abdominal nerves, glands, and vessels generally.

Veratrin exercises a specific influence over the uterus, and has been beneficially employed in amenorrhea, atonic chlorosis, uterine leucorrhea, and other affections dependent upon vascular debility.

Veratrin is valuable in the treatment of jaundice, when arising from obstructions of the liver and portal circulation. Also in dropsical affections characterised by much coldness and torpor, or when arising from suppression of the catamenia or hemorrhoidal flux. In all diseases of the mucous membranes of the intestines, phlegmatic obstructions, tympanites, fleshy tumors of the abdomen, chronic enlargment of the liver, spleen, and mesentery, and in debility of the muscular fibres of the intestinal tube, the Veratrin is a valuable remedy. In these disorders, it operates most beneficially in connection with Podophyllin. In cachexies and dyscrasies, arising from functional aberation of the viscera of the abdomen, particularly of the liver and mesenteric glands, and in herpes, and other diseases of the skin, Veratrin is highly recommended. Also in atrabilious, arthritic, and rheumatic dyscrasies.

Experience seems to prove that a majority of the cases of mental aberation, and of nervous diseases generally, arise from and are dependent upon a morbidly increased activity of the nervous structure of the abdomen, functional obstructions and organic lesions of the abdominal viscera, and disturbed and discordant action of the abdominal nervous plexus. At any rate, if such be not the case, experience proves that such remedies as act as stimulants and alteratives upon the nervous tissues of the abdomen, are most beneficial in that class of diseases. This would seem to explain why Veratrin is of utility in the treatment of mania, epilepsy, hysteric cramps, chronic convulsions, melancholy, and mental weakness. Certain it is that when material obstructions are ascertained to exist, particularly of the sanguineous secretions, as the catamenia, hemorrhoids, etc., relief is almost certain to follow the exhibition of the Veratrin. In cases of mental aberation accompanied with torpor and debility of the abdominal organs, Veratrin will be found serviceable. In such cases it may be administered in full doses, even to the production of an emeto-cathartic effect, observing much caution, however, in its exhibition. Where visceral obstructions are of long standing, the treatment should be premised with the judicious use of Podophyllin, which will materially enhance the efficacy of Veratrin in all cases in which the former may be indicated.

In Chronic pneumonic and catarrhal affections, having a tendency to effusion and exudation, and in chronic rheumatic affections of the lungs and pleura, and which are so often connected with hydrothorax, Veratrin is an excellent remedy. In these cases it may sometimes be advantageously joined with Digitalin, and alternated with suitable doses of a combination of Podophyllin, Asclepin, and Cerasein.

Veratrin is also of much value in the treatment of some forms of scrofula, particularly when occurring in persons laboring under vascular repletion, and whose lymphatic system is in an inactive or torpid condition.

Veratrin is contra-indicated in all cases of paralytic debility, tendency to hermorhage of the lungs, pregnancy, lingering hectic, internal ulcerations, etc.

The dose of Veratrin will vary, according to the impressibility of the patient's system, and the requirements of the case. In general this variation will be from ONE-EIGHTH to ONE-HALF of ONE GRAIN. In febrile forms of disease, small doses, frequently repeated, are of most service; while in chronic affections, as in disease of the heart, dropsies, etc., larger doses, and at longer intervals, are preferable. In fevers, we usually administer it every two hours. In chronic affections of the mucous membranes, visceral engorgements, etc., twice per day. In cardiac diseases, whenever the urgent symptoms arise. In croup, convulsions, and asthma, at intervals of thirty minutes, until the spasm is broken, and relief afforded, and then with such frequency of repetition as may be necessary to maintain the desired influence. To what extent the patient may be subjected to its influence, or for what length of time this influence may be continued, without danger to the patient, is a question difficult of solution. A case of scarlatina in a girl some ten or twelve years old came under our observation, in which the patient was kept so completely under its control that, for the period of forty-eight hours, no pulse could be felt at the wrist. At the end of this time, the fever having been subdued, the medicine was omitted, the circulation rose to the normal standard, and the patient had a rapid convalescence. It was a most malignant case, and one which, when the treatment was commenced, afforded little prospect of recovery. It is fair to state that the preparation employed in this case was the concentrated tincture.

It is of the highest importance, under all circumstances of the employment of Veratrin, to previously neutralise undue acidity of the stomach, and to administer it in such form as to render it most diffusible. We generally prefer to triturate it with Asclepin, and to administer it in solution. We find very few cases in which the Asclepin is contra-indicated.

Except the above, very few judicious combinations can be effected with the Veratrin. In some forms of cardiac disease, and in dropsical effusions, it may be beneficially joined with Digitalin, as previously mentioned. When auxiliary remedies are needed, we deem our practice of alternation the best. Care must be exercised when employing the Veratrin in chronic diseases, in order that too great a degree of relaxation and prostration be not produced, which must be obviated by alternating with suitable stimulants and tonics.

# CON. TINC. VERATRUM VIRIDE.

Derivation same as the Veratrin. The properties and appliances of this tincture are the same in all respects as those of the Veratrin. Its relative medicinal strength is as EIGHT to ONE. That is, EIGHT DROPS of the tincture are equivalent to ONE GRAIN of Veratrin. We prefer it to the Veratrin on account of its advantage of ready administration, as well as on account of its diffusible character. In view of the latter quality, we consider it more prompt in its influences than the Veratrin. We use it almost exclusively in our practice. The average dose, as a diaphoretic, anti-spasmodic, febrifuge, and arterial sedative, is TWO drops, repeated once in two hours. As an emetic, in croup, convulsions, etc., from FIVE to EIGHT drops, repeated every one or two hours. In chronic diseases generally, we give from ONE to THREE drops thrice a day. In asthma and affections of the heart, we generally prescribe it when the urgent symptoms are present. The most convenient form for administration in fevers and other acute diseases, is as follows:

℞.
  Asclepin............................ ℨ ss.
  Warm water......................... ℥ IV.
  Con. Tinc. Veratrum............... gtt. XXX.

Dissolve the Asclepin in the water and add the Veratrum.

Stir the solution well when used. Dose, from one to three teaspoonfuls once in two hours. If nausea arises, and vomiting be not desirable, omit the medicine until it subsides, and then resume in the same manner, or at longer intervals.

In this form we employ it in remittent, scarlet, and typhoid fevers, pneumonia, pleuritis, measles, acute rheumatism, dysentery, all forms of acute exanthema, and febrile diseases generally. In cardiac affections, and in dropsies, it may be combined with the Con. Tinc. Digitalis in equal proportions.

The Con. Tinc. of Veratrum has been found a most excellent external application for the relief of neuralgia and rheumatic pains, and for the discussion of indolent scrofulous and other tumors, enlarged glands &c. The parts may be bathed with the tincture two or three times per day, or a cloth saturated with it may be bound upon the tumor or part affected.

We have been informed that the tincture has been successfully employed, in enemas, for the removal of the ascaris vermicularis or pin worms of the rectum, but of this fact we have no personal knowledge. From FIVE to TEN drops may be administered in from TWO to FOUR ounces of water. We should prefer a thin mucilage of slippery elm, or a solution of molasses and water. It is worthy of further trial in this respect. Much yet remains to be learned of the value of Veratrum and its preparations, although sufficient is already known to render it an indispensible agent in the hands of every practitioner. Its positive yet kindly control over the heart and arterial system, by means of which we may say to the turbulent currents of the blood, with certainty of obedience, "Peace, be still," constitutes it a *sine qua non* in the treatment of febrile diseases. In addition, its power of resolving the plasticity of the blood, its stimulant, alterative, and tonic influences over the venous, absorbent, and lymphatic vessels and glands, and its power of promoting the sanguineous secretions, renders it of inestimable utility to the requirements of the healing art. We would wish especially to note an important fact in connection with the employment of this remedy in the treatment of acute diseases, and that is, we can truly pronounce our patient *well* when he is discharged. No ptyalism—no loosening of the teeth—no sloughing of the soft parts—no lesions of the mucous membranes or other tissues—no morbid discharges from the eyes or ears, as is frequently the case in scarlatina and measles—no troublesome eczema to harass the weary sufferer—no barometric pains to announce approaching meteoric change—nor fetid ulcers discharging their filthy ooze from fountains of corrupt and stagnated secretions within:—but a system renovated and invigorated—the vital currents leaping in living joy through their unobstructed channels—the unfettered nerves harmoniously obedient to the mandates of the organic intelligence, and the rose of health blooming in grateful acknowledgment over the integrity of the soul's citadel.

# EUPATORIN. [Purpu.]

---

Derived from *Eupatorium Purpureum*.
Nat. Ord.—*Asteraceæ*.
Sex. Syst.—*Syngenesia Æqualis*.
Common Names.—*Queen of the meadow, Gravel weed, Joe-pye, Trumpet weed*, etc,
Part Used.—*The Root*.
No. of Principles.—*three*, viz., *resinoid, neutral and alkaloid*.
Properties.—*Diuretic, stimulant, astringent and tonic*.
Employment.—*Gravel, dropsy, gout, rheumatism, hematuria, hematamesis, hemoptysis, dysentery, hooping cough, asthma*, etc.

ALTHOUGH the system of nomenclature adopted by the manufacturers of concentrated medicines is calculated to create some confusion when two or more plants are taken from the same genera, we deem it better, until a uniform system of terminology is devised, to designate the preparations of plants belonging to the same genus by an abbreviation of their specific names, as in the instance before us. Preparations purporting to be the active principles of this plant have been offered the profession under the designation of *Eupurpurin*, etc., but we are at a loss to discover either scientific authority, or advantage in the name adopted. We shall, therefore, for the present, adhere to the method of distinction herein pursued, as the preparation of which we are treating has already been introduced to the profession under the title above given.

The value of Eupatorin Purpu. in the treatment of gravelly affections depends more upon its alterative than upon its direct diuretic influences. It seems more effectual in the removal of uric acid deposits than of other calculous formations, although it is beneficially employed in almost all affections of the kidneys and bladder. It resolves mucous deposits and deterges and heals abraded mucous surfaces. In catarrh of the bladder, engorgement of the ureters, and in all atonic conditions of the urinary apparatus, it is peculiarly useful. In dropsy, strangury, hematuria, gout and rheumatism, it is a valuable auxiliary agent. Its utility in the last mentioned diseases is owing to the power of resolving the viscidity of the secretions, and of promoting renal depuration. The average dose, in chronic disorders, is THREE grains three times per day. The quantity may sometimes be increased to FIVE, and even TEN grains, with safety and advantage. In acute affections, as hematuria, strangury, etc., the doses may be repeated every one or two hours. Its efficacy will be enhanced and its action rendered more prompt, in these cases, by administering it in solution in warm water.

Eupatorin Purpu. operates beneficially in dropsy by reason of its stimulating influence upon the absorbent vessels, as well as by its powers as a diuretic. In this complaint it may be joined with Ampelopsin, Helonin, and other of the concentrated remedies, as mentioned under their respective heads.

Hemoptysis, hematamesis, and other hemorrhages, have been arrested and cured by the use of this remedy. The doses in these cases will vary from TWO to FIVE grains every thirty or sixty minutes, or at longer intervals, according to the urgency of the symptoms. If desired, it may be combined with Lycopin, or Geranin, or Trilliin, or Myricin, etc We have found it of great value in dysentery, both as an astringent, when such is needed, and as a tonic in the convalescing stages. It seems to exercise a peculiar soothing and toning influence upon inflamed and abraded mucous surfaces. It promotes assimilation, and restrains the diarrheal tendency.

We have also found the Eupatorin Purpu. a most excellent remedy in whooping cough, asthma, and other affections of the respiratory system. We set a high value upon it as an expectorant. It resolves the viscidity of the pulmonary secretions, resolves the plasticity of the venous blood, and promotes cutaneous depuration. We are also inclined to attribute much of its efficacy in these affections to its influences as an alterative and diuretic upon the urinary apparatus, as we believe that many cases of apparent disease of the lungs are dependent upon the retention of effete urinary materials. At any rate we have frequently found diuretics to be the best remedies in whooping cough, asthma, and chronic coughs generally. And in the treatment of dermoid diseases, we class those alteratives possessing diuretic properties as the most efficient in the materia medica. It is for the reason above given that we employ this remedy in the diseases above mentioned, and we find it efficient and reliable. If the patient partake of warm diluent drinks in connection with the Eupatorin, a mild and pleasant diaphoresis is produced.

We have found this remedy beneficial in all cases of dyspnea, no matter by what cause produced. Also in catarrh, influenza, bronchitis, and phthisis.

# EUPATORIN, (Perfo.)

Derived from *Eupatorium Perfoliatum.*
Nat. Ord.—*Asteraceœ.*
Sex. Syst.—*Syngensia Æqualis.*
Common Names.—*Boneset, Thoroughwort, etc.*
Part Used.—*The Herb.*
No. of Principles—*three,* viz., *resinoid, neutral,* and *alkaloid.*
Properties.—*Aperient, emetic, diaphoretic, febrifuge, alterative, resolvent,* and *tonic.*
Employment.—*Intermittent, remittent, typhoid,* and other *fevers, coughs, colds, influenza, catarrh, dyspepsia, debility, etc.*

EUPATORIN is alterative, resolvent, tonic and aperient when taken in small doses and administered in powder or pill; and emetic, diaphoretic and febrifuge when exhibited in a warm fluid menstruum. Hence the form of its administration will be governed by the necessities of the case. It is much employed, in solution in warm water, to facilitate the operation of other emetics. It is a valuable diaphoretic and febrifuge in all febrile diseases, when given in small and frequently repeated doses. Intermittent and remittent fevers have been effectually cured by administering the Eupatorin in full emetic doses during the intermissions or remissions, and as near the time of the expected chill or exacerbation as possible, following with small repeated doses to the production of free diaphoresis, which should be continued uninterruptedly for six or eight hours, and then employing the remedy in cold solution, pill, or powder, as a tonic. In consequence of its utility in periodic fevers, Eupatorin has been accredited with anti-periodic powers. We are of opinion, however, that this property no more pertains to it than to tonics in general. In all fevers and other affections manifesting a tendency to putrescency of the fluids, Eupatorin has been found of excellent service, seeming to exercise well marked and desirable influences as an antiseptic. Hence it is employed in typhoid and typhus fevers, epidemic dysentery, erysipelas, putrid sore throat, etc.

Eupatorin has been found useful in chronic cough, senile debility, constipation, diseases of the skin, loss of appetite, languid circulation, whooping cough, asthma, etc.

The dose of the Eupatorin as an emetic is from FIVE to TEN grains in warm water, repeated every thirty minutes until it operates. As an emetic it is slow but thorough. When given in full doses it generally acts upon the bowels. It is valuable, in warm solution, for promoting the operation of other emetics.

As a diaphoretic and febrifuge, from ONE to THREE grains may be given once in two hours, in warm water, or in an infusion of some aromatic herb, as catnep, pennyroyal, spearmint, etc. It may be joined with other diaphoretics, as the Asclepin, Cypripedin, or Sanguinarin.

As a tonic and aperient, from THREE to FIVE grains three times a day, in cold water, pill, or syrup. It is also a valuable alterative or resolvent, useful in scrofulous and other cachexies, tinea capitis, eczema, herpes, and other cutaneous diseases

# .CON. TINC. EUPATORIUM PURPUREUM.

This tincture may be used for all the purposes of the preceding preparation. The average dose is SIX drops, increased or diminished as occasion requires. When desired, it may be combined with other of the concentrated tinctures. Thus in dysentery, intestinal ulcerations, etc., with the Con. Tinc. Rhus Glab., as follows:

℞.
    Con. Tinc. Eupatorium Purpu. ........... ℨij.
    "   " Rhus Glab. .................... ℨj.

Mix. Dose from SIX to TEN drops. In the asthenic forms of dropsy with Con. Tinc. Euonymus:

℞.
    Con. Tinc. Eupatoriu Purpum............ ℨj.
    "   " Euonymus.................... ℨss.

Mix. Dose from FOUR to EIGHT drops, or more. In this way various combinations may be effected suited to the case in hand. It is convenient of administration, and appropriate in hematuria and other cases in which promptitude of action is desirable.

# CORYDALIN.

---

Derived from *Corydalis Formosa*.
Nat. Ord.—*Fumariaceæ*.
Sex. Syst.—*Diadelphia Hexandria*.
Common Names.—*Turkey Corn, Turkey Pea, Staggerweed, etc.*
Part Used.—*The Root*.
No. of Principles, *four*, viz., *resin, resinoid, alkaloid,* and *neutral*.
Properties.—*Alterative, tonic, diuretic, anti-syphilitic, antiscorbutic, resolvent, etc.*
Employment.—*Scrofula, syphilis, cutaneous diseases, dropsy, debility, etc.*

This plant is the *Dielytra Eximia* of Wood's, and the *Diecentra Eximia* of Gray's botany.

The remedial properties of this plant are of a very high order, and reside, as above stated, in four distinct proximate active principles. This combination of the active medicinal constituents embodies the entire therapeutic value of the plant. Our clinical experience in the use of both the crude root and its concentrated preparation enables us to speak positively upon this point.

The therapeutic action of the Corydalin is at once both remarkable and highly to be prized. With the most energetic alterative and resolvent properties, it combines a tonic power of exceeding value. Thus while it neutralises, deterges, and promotes depuration, it gives tone to the various organs engaged in the performance of these functions. Its dynamic influences seem to be comprised in a power by which it resolves the plasticity of the blood, regulates and quickens the activity of the eliminating vessels, particularly of the renal and cutaneous, and promotes the processes of digestion, assimilation, and nutrition. From this consideration of its physiological influences, it will be at once seen that the Corydalin admits of an extended and desirable range of application. In scrofula, particularly when accompanied with feeble digestion and poverty of the blood, it is of great value. As this disease almost invariably argues an atonic condition of the reparative and depurative functions, the peculiar efficacy of the Corydalin will be apparent. In this complaint the Corydalin should be given in doses of from ONE to THREE grains three times per day, alternating with such other remedies as may be needed to correct hepatic aberation or other special visceral derangements. The practitioner may combine it, when he deems it expedient, with other alteratives, diuretics, or tonics, as Senecin, Ampelopsin, Cerasein, Stillingin, Irisin, etc. We prefer, however, to alternate it with such other remedies as the necessities of the case may indicate.

Corydalin has been employed with marked success in the treatment of syphilis, in connection with Podophyllin. Perhaps no single remedial agent possesses more positive and energetic anti-syphiltic and anti-scorbutic properties. Its use should be persevered in for a length of time, occasionally alternated with Stillingin, Phytolacin, Irisin, etc. The most desperate and protracted cases have been cured by this treatment.

Our experience in the use of this remedy in the treatment

of cutaneous eruptions has been highly satisfactory. We have succeeded in curing many cases of obstinate dermoid affections, when other remedies proved inefficient, by the use of the Corydalin. On account of the smallness of the dose and the absence of any nauseous taste, it is peculiarly adapted to the necessities of children. It may be readily administered in solution, in a little water. When not contra-indicated, a little sugar may be added, which will render it of easy administration to infants and children. In strumous, herpetic, venereal, scorbutic, and other cachexies, the Corydalin is worthy the entire confidence of the profession.

Corydalin is also valuable in dropsy, general debility, gravel, and the various affections of the urinary apparatus, indigestion, torpor of the lacteals, visceral enlargements, and for the correction of all vitiated conditions of the blood and fluids.

As a diuretic, the Corydalin is more to be valued on account of its resolvent and alterative properties than for its direct influence in increasing the secretion of urine. In atonic gleet, passive leucorrhea, catarrhal affections of the bladder, incontinence of urine, etc., it will be found peculiarly serviceable.

The average dose of the Corydalin is TWO grains. It seldom or never disagrees with the stomach, and may be employed as a tonic in irritable conditions of that organ.

# JUGLANDIN.

Derived from *Juglans Cinerea.*
Nat. Ord.—*Juglandaceæ.*
Sex. Syst.—*Monœcia Polyandria.*
Common Names.—*Butternut, White Walnut, etc.*
Part Used.—*Bark of the Root.*
No. of Principles.—*Two,* viz., *resinoid* and *neutral.*

Properties.—*Alterative, tonic, chologogue, laxative, deobstruent, detergent and diuretic, and in large doses emetic and cathartic.*

Employment.—*Fevers, dysentery, dyspepsia, piles, jaundice, hepatic disorders, and diseases of the urinary apparatus.*

JUGLANDIN is a remedy of great value. As a laxative and cathartic, it is devoid of irritant properties, hence is exceedingly useful in all forms of bowel complaints, and in fevers and other disorders attended with gastric or enteric irritability, when such a remedy is indicated. We have employed the Juglandin with much satisfaction in the treatment of intermittent, remittent, and typhoid fevers accompanied with gastric irritability and a tendency to diarrhea. It corrects the acrimony of the secretions, neutralises acidity, obviates the tendency to fermentative decomposition of the food, stimulates the hepatic secretions, resolves biliary deposits, deterges and soothes the irritability of the mucous surfaces, promotes peristaltic activity, and gives tone to the depurative functions of the kidneys. From this statement of its capabilities, it will be seen that its range of application is extensive.

The average dose of the Juglandin is FIVE grains. In large doses, say from TEN to FIFTEEN grains, it generally proves cathartic, and sometimes emetic, accompanied with vomiting of bilious matter. It is as an aperient and laxative, however, that the Juglandin is mostly esteemed, its cathartic powers being somewhat uncertain.

In indigestion accompanied with gastric irritability, flatulency, acid eructations, etc., we have employed the Juglandin with the most gratifying success. We usually administer it in doses of FIVE grains immediately after each meal. We imbibed a notion, some years since, that medicines calculated to excite action in the digestive apparatus should be so administered as to expend their influences at the moment when such action was needed, and our experience has fully justified us in the correctness of the opinion then formed. The benefit here derived results from a local influence, hence by so timing the remedy that it may promote the action called forth by the natural excitant, food, we secure the benefits of its co-operation. If the muscular fibre be lax and inactive, its contractile powers are stimulated into activity, and thus is the labor of attrition promoted. If the gastric secretions be deficient, dependent upon atony or torpor of the gastric functions, they are incited to yield up their stores of the digestive juices. Incorporated with the chyme as it passes into the intestinal tube, the medicinal constituents provoke a due supply of bile and pancreatic juice, flow onward with the duly elaborated chyle, quicken the impressibility of the lacteal vessels, and impel the life-sustaining currents forward to the completion of their organic mission. We are further of the opinion that certain chemical relations are sustained, whereby the processes of assimilation are facilitated. Be these considerations as they may, however, our plan of administering medicines calculated to act as local stimulants and tonics will be found reliable.

Juglandin, answers an admirable purpose in combination with Leptandrin. Our formula is as follows:

℞.
    Juglandin  
    Leptandrin     -   -   -   -    aa. ʒj.  
    Mucilage Gum Acacia   -  -   q. s.

Form a mass and divide into THIRTY pills.

We have found these pills to answer an excellent purpose in the treatment of indigestion, chronic hepatic disorders, constipation, jaundice, piles, and derangements of the urinary apparatus. The usual dose is ONE pill, taken immediately before or after each meal. If necessary, to obviate constipation, the dose may be increased to TWO or THREE pills, or from TWO to FOUR may be taken at bed time. We are confident that whoever tests the value of these pills will never be without a supply of them on hand. They correct a tendency to fermentative decomposition of the food, deterge and soothe the irritability of the mucous membranes, obviate constipation, expel flatulence, and correct the acrimony of the urine. In atonic conditions of the stomach and bowels, and in general debility and torpor of the abdominal viscera, we substitute the Con. Tinc. Xanthoxylum for the mucilage of gum arabic in forming a mass for pills. When a milder stimulant is needed, we employ the Xanthoxylin, which, being deprived of the oil, is not incompatible in conditions of sub-acute inflammation. Our formula then stands as follows:

℞.
    Juglandin  
    Leptandrin  - - - - -  aa. ʒss.  
    Xanthoxylin  - - - -  ℈i.  
    Mucil. acacia  - - - - -  q. s.

Form a mass and divide into TWENTY pills. Dose, same as above.

For the relief of ischuria, eneuresis, and kindred disorders of the urinary apparatus, the Juglandin will operate most efficiently in combination with Populin  They may be combined in equal proportions and formed into FOUR grain pills, one of which may be given every two hours, or oftener, until relief is obtained, and then continued at suitable intervals until a cure is effected. These will be found excellent for the relief of scalding of the urine in pregnant females, and in the treatment of cystitis and urethral inflammation.

In dysentery, the Juglandin is usually administered at intervals of two hours, and continued until the alvine discharges assume a healthier appearance. The average dose in such cases is TWO grains. When indicated, it may be alternate with Geranin or other astringents.

# TRILLIIN.

Derived from *Trillium Pendulum.*
Nat. Order.—*Trilliaceæ.*
Sex. Syst.—*Hexandria. Trigynia.*
Common Names.—*Beth-root, Birth-root, etc.*
Part Used.—The *Root.*
No. of Principles.—*Three*, viz., *resinoid, neutral* and *muci-resin.*

Properties.—*Astringent, styptic, alterative, tonic, diaphoretic, expectorant, anti-septic* and *emmenagogue.*

Employment.—*Hemorrhages*, either *external* or *internal, leucorrhea, prolapsus uteri, menorrhagia, dyspepsia, hooping cough, asthma, immoderate flow of the lochia,* etc.

TRILLIIN is one of the most valuable agents embodied in the organic materia medica. Its dynamic influences are chiefly directed towards the mucous surfaces, over which it seems to exercise a special control. Though mostly employed in affections of the uterine system, it is nevertheless of great utility in the treatment of all diseases involving the mucous membranes. Hemoptysis, hematemesis, hematuria, and uterine hemorrhages have all been relieved and cured by means of this remedy. The average dose in these cases is THREE grains, repeated hourly until the hemorrhage is arrested, and then continued at intervals of from four to six hours until all danger of a relapse is past. Relief will be rendered more certain if the Trilliin be alternated with Oil of Erigeron, FIVE drops of which may be given every alternate hour. Or it may be alternated with Lycopin, of which from TWO to FOUR grains may be given at a dose. In chronic cough, accompanied with spitting of blood, the Trilliin and Lycopin may be combined, as follows:

℞.
Trilliin............................
Lycopin............................ ℨss.

Mix and divide into FIFTEEN powders. Dose, one, three times a day. This combination will also be found excellent in diabetes, and, in connection with suitable diet and regimen, will be found successful in a majority of cases, if taken in the early stages.

In the treatment of vaginal and uterine leucorrhea, particularly when of an atonic character, the Trilliin will be found one of the most reliable remedies. It resolves the viscidity of the mucous secretions, acts as an alterative tonic upon the mucous follicles, deterges and heals the diseased membranes, and corrects the acrimony of the discharges. Trilliin is decidedly antiseptic, and is useful in correcting a tendency to putrescency of the fluids, and the foetor of critical discharges. In dysentery, putrid fevers, cancrum oris, and in all cases having a tendency to gangrene, it will be found of essential service. When required, it may be applied locally, either in the form of a solution, as in cancrum oris, putrid sore throat, etc., as a gargle, or the dry powder may be applied, as in erysipelatous and other ulcers. In fetid discharges from the vagina and uterus, it may be employed in the form of an injection. For this purpose, from ONE to TWO drachms may be infused in boiling water and used when blood warm. For the latter purpose it may be combined with Geranin, or Myricin, or Baptinin, ONE drachm of each to the pint. Thus combined it will be found useful as an injection in vaginal, uterine, and rectal hemorrhages. A solution of the Trilliin, or a small quantity of the dry powder, snuffed up the nostrils will immediately check epistaxis. A small quantity of the powder introduced into the cavity from which a tooth has been extracted will effectually arrest the hemorrhage. Slight hemorrhages occurring from wounds, cancerous ulcerations, etc., may also be arrested by the same means.

But among the most valuable of the hæmostatic properties of the Trilliin is its power of restraining profuse lochial discharges. It facilitates the detergent action, regulating but not suppressing it. It may be given in doses of from TWO to FOUR grains three times a day, or oftener if the indications warrant.

We have also found the Trilliin exceedingly valuable in the treatment of prolapsus uteri, particularly when of an asthenic character, and dependent upon an atonic condition of the uterine supports. It should be given in doses of from TWO to FIVE grains three times per day, and alternated with such other remedies as the case demands. In engorgements of the cervix uteri, chronic vaginitis, etc., the Trilliin will be found an exceedingly efficient remedy, and should be used both internally and externally. In passive hemorrhages of the uterus and other organs, the Trilliin, if not sufficient alone, will always prove a valuable auxiliary.

Trilliin has been highly recommended in dyspepsia, hooping cough, asthma, etc., and we have no doubt of its utility in these complaints, although our personal experience of its value in such cases is too limited to allow us to speak authoratively. The average dose of the Trillin is THREE grains.

# SCUTELLARIN.

**Derived** from *Scutellaria Lateriflora.*
Nat. Ord.—*Laminaceæ.*
Sex. Syst.—*Didynamia Gymnosperma.*
Common Names.—*Blue Sculloap, Mad Dog Weed,* etc.
Part Used.—The *Herb.*
No. of Principles—*three,* viz., *resin, resinoid,* and *neutral.*
Properties.—*Nervine, tonic, diuretic, and anti-spasmodic.*
Employment.—*Convulsions, chorea, delirium, hysteria, dysmenorrhea, neuralgia, nervous debility, urinary disorders,* etc.

MUCH division of sentiment has heretofore existed among the profession in regard to the remedial value of the Scutellaria Lateriflora. By many it is considered a medicine of great utility in the treatment of a variety of disorders, while others attach little or no value to it. We have shown, in the first part of this volume, while treating of the variations in the therapeutic constituents of plants, that this discrepancy of opinion had good foundation, in view of the different degrees of development attained by the proximate active principles under diverse local influences. The presence of a greater or less amount, or the entire absence of those constituents upon which a plant depends for medicinal value must ever give rise to a division of sentiment respecting its claims as a therapeutic agent. Whoever uses the Scutellarin now being treated of, will not fail to place it in his catalogue of remedies as a medicine entitled to his confidence. An ounce of Scutellarin being positive and uniform in its constitution and properties, will better enable him to determine its worth than a thousand pounds of the crude herb.

As a nervine tonic, we value the Scutellarin highly. It soothes and quiets the irritability of the nervous system, giving tone and regularity of action, lessens cerebral excitement, abates delirium, diminishes febrile excitement, excites diaphoresis and diuresis, and accomplishes its work without any subsequent unpleasant reactions. The average dose of the Scutellarin is TWO grains, increased, when occasion requires, to FIVE, or even more. The doses may be repeated every one, two, or three hours, according to the urgency of the symptoms.

Scutellarin is of great service in fevers and other acute diseases in which there is a tendency to delirium. It seems to have the power of lessening cerebral excitement, and at the same time proves febrifuge. It is equally useful in the treatment of acute dysmenorrhea, menorrhagia, and other female disorders in which the head is liable to be unpleasantly affected. It would seem to have an especial influence in equalising the flow of the nervous currents, and so lessening the tendency to congestions. We have found the Scutellarin a remedy of great value in the treatment of *coup de soleil* or sun stroke, particularly when the case has become chronic. We meet with many patients who have been unpleasantly affected by heat, and who have never entirely recovered from its effects. They are unable to endure the sun's rays, and complain of dizziness, headache, nervous tremblings, wakefulness, indigestion, etc. We have met with entire success in many of these cases by the use of the Scutellarin in connection with Podophyllin. We administer the latter in full cathartic doses at the commencement of the treatment, in view of its derivative influences, and afterwards repeat it in such doses and at such intervals as in our judgment may be necessary. The Scutellarin we exhibit in doses of from TWO to FIVE grains three or four times a day.

Scutellarin is an excellent remedy in the treatment of convulsions, chorea, hysteria, etc., more as a radical remedy during the remissions, however, than as a means of overcoming the immediate spasm. It seems to be of more utility, in these cases, as a means of giving permanency to a condition, than as a means of bringing about a condition. In the treatment of epileptic convulsions, as soon as we have secured a remission of the attacks by means of Gelsemin, we employ the Scutellarin in combination with the Gelsemin as a radical remedy. We find them to operate admirably in combination Our formula is as follows:

℞.
    Scutellarin .......................... ℈ i.
    Gelsemin............................grs. V.

Mix and divide into ten powders. Dose, one, two or three times a day. The proportions may be varied to suit the peculiarities of the case.

Scutellarin may be relied upon under all circumstances as a nervine tonic. In all cases of nervous irritability, debility, hysteria, dysmenorrhea, etc., indications will be found for its employment. We deem it much superior to the preparations of opium in the management of the disorders of children. For nervous irritability, wakefulness, slight febrile disturbances, flatulence, colicky pains, etc., it answers admirably in combination with Asclepin. Make a solution in warm water and administer in small and frequently repeated doses.

When desired, the Scutellarin may be combined with other antispasmodics, as the Caulophyllin, Viburnin, Cypripedin, etc., or with tonics, as the Cerasein, Cornin, Fraserin, etc., or with diuretics, as the Populin, Senecin, Eupatorin Purpu., etc.

We have used the Scutellarin with benefit in threatened trismus, tetanic cramps, and other spasmodic disorders. Its diuretic powers are considerable, but not uniformly displayed.

In many cases we have found it to induce a copious flow of urine, while in others no appreciable diuretic effects were observable. When taken in warm solution it proves gently diaphoretic, and is useful in breaking up a recent cold.

# CON. TINC. SCUTELLARIA LATERIFLORA

This preparation of the Scutellaria is equivalent to the preparation first treated of, and is employed for the same purposes. It is convenient of dispensation and administration, and for combining with other of the Concentrated Tinctures. We make much use of it in combination with the Con. Tinc. Gelseminum.

℞.
    Con. Tinc. Scutellaria................
    "   "  Gelseminum................aa. ℨ ii.

Mix. Dose from FIVE to FIFTEEN drops. We employ it in epileptic and other convulsions, hysteria, dysmenorrhea, chorea, nervous debility, wakefulness, etc.

Equal parts of the Con. Tinctures of Scutellaria and Senecio form an excellent combination for the treatment of pectoral disorders, gravelly affections, amenorrhea, nervous debility, hysteria, uterine engorgements, and other disorders of the female system.

With the Con. Tinc. Eupatorium Purpu. it will be found serviceable in affections of the urinary apparatus.

Either alone or in combination with other suitable agents. it will be found valuable for the relief of nervous headaches, neuralgic pains, palpitation of the heart, and in all disorders indicating the employment of an antispasmodic, nervine, tonic, and diuretic.

The average dose of the tincture is FIVE drops, varied as circumstances may require. It produces no unpleasant effects in over doses, operating under all circumstances, so far as we have observed, without excitement.

# OIL OF POPULUS TREMULOIDES.

WE omitted to notice the Oil of Populus in its proper connection, hence introduce it here. It is chiefly as an external application that we desire to call attention to it, its value as an internal remedy being so indefinite that we prefer omitting any reference to its internal employment. As an external appliance for burns, sore nipples, abrasions of the skin, and various eruptions, we are enabled to speak from experience of its great value. In its influences it seems to partake of the character of the balsams, soothing irritation, correcting the acrimony of eruptive exudations, and favoring cicatrization. For the purposes above mentioned it may be made into an ointment with lard, fresh butter, simple cerate, or other bases. From ONE to THREE drachms of the Oil may be added to each ounce of the base employed. In some cases it may be usefully joined with Olive oil, or oil of sweet almonds. The same proportions above mentioned may be observed. At other times it may be requisite to apply the oil without admixture. For some forms of eczema, salt rheum, excoriated nipples, burns, scalds, abrasions, healthy ulcers, etc., this will be found one of the most efficient applications ever employed. In eruptions of the scalp it will be found equally useful.

# APOCYNIN.

---

Derived from *Apocynum Cannabinum.*
Nat. Ord.—*Apocynaceæ.*
Sex. Syst.—*Pentandria Digynia.*
Common Names.—*Black Indian Hemp, Dog's-bane, etc.*
Part Used.—*The Root.*
No. of Principles, *three*, viz., *resin, resinoid,* and *neutral,*
Properties.—*Emetic, cathartic, diuretic, diaphoretic, alterative, tonic, and vermifuge.*

Employment.—*Intermittent and remittent fevers, rheumatism, scrofula, dropsy, syphilis, constipation, chronic hepatitis, jaundice, etc.*

IN small doses, say from ONE FOURTH to ONE HALF of ONE grain, Apocynin is diaphoretic, expectorant, stimulant, and diuretic, and as such is employed in intermittent and remittent fevers, pneumonia, pleuritis, acute rheumatism, and other febrile disorders. In large doses it is an active emeto-cathartic, somewhat drastic in its operation, producing copious watery stools, and greatly promoting diuresis.

We have found the Apocynin efficient in promoting the

absorption of serous effusions, particularly when investing the larger cavities, as of the chest, abdomen, etc. We have employed it with success in the treatment of hydrothorax. The average dose of the Apocynin is TWO grains, repeated twice or thrice daily. It frequently produces considerable nausea and griping, which may be corrected by combining it with aromatics and stimulants. As a diaphoretic and expectorant, from ONE FOURTH to ONE HALF GRAIN may be given once in from two to four hours. It seems to resolve the viscidity of the pulmonary secretions, and to stimulate the mucous surfaces into healthful activity, hence is useful in bronchitis, laryngitis, catarrh, etc. We have employed the Apocynin successfully in the treatment of hemoptysis. It is most useful when the latter results from the suppression of some secretion, as the menses, hemorrhoids, or from serous accumulations within the cavity of the chest.

Apocynin has also been found serviceable in scrofula, syphilis, eruptions of the skin, constipation of the bowels, chronic hepatic aberation, jaundice, and for the removal of worms, particularly the *ascaris vermicularis*. For the latter purpose it is administered three times daily, in doses sufficient to keep the bowels somewhat relaxed, continued for three days, then omitted for three days, and resumed again if required.

Apocynin is accredited with some narcotic power, in view of the patient's becoming somewhat drowsy when under the influence of cathartic doses. The pulse at the same time diminishes in frequency. These effects pass off, however, with the operation of the medicine. When given in too large, or too frequently repeated doses, a lingering and distressing nausea is produced, accompanied with prostration and debility. In the treatment of scrofula and other diseases of an asthenic character, it should be alternated with tonics. Combinations with other remedies may be easily effected, at the option of the practitioner, but we are decidedly in favor of using it singly and alternating with other remedies when indicated.

The diuretic power of the Apocynin seems to reside more

in its property as a stimulant of the absorbent system, than in any direct influence it has upon the kidneys. For this reason it will be observed that its operation as a diuretic is not uniform, and is governed by the existing diathesis.

## CON. TINC. APOCYNUM CANNABINUM.

DERIVATION, properties and employment same as the above. Medium dose, THREE drops. Preferred by many on account of its diffusible character, and the facility with which it may be administered. When desired, it may be combined with other of the concentrated tinctures indicated in the case. The following combinations are sometimes employed:

℞.
    Con. Tinc. Apocynum..................
    "   "   Chelone....................aa. ʒ i.
Mix. Dose from TWO to FIVE drops.

℞.
    Con. Tinc. Apocynum..................
    "   "   Euonymus..................aa. ʒ i.
Mix. Dose same as above.

℞.
    Con. Tinc. Apocynum .................... ʒ i.
    "   "   Eupatorium Purpu............ ʒ ii.
Mix. Dose from FOUR to EIGHT drops.

# BAROSMIN.

Derived from *Barosma Crenata*.
Nat. Ord.—*Rutaceæ*.
Sex. Syst.—*Pentandria Monogynia*.
Common Name.—*Buchu*.
Part Used.—*The Leaves*.
No. of Principles.—Two (*Resin and Neutral*.)

Properties.—*Diuretic, alterative, diaphoretic, tonic, stimulant and antispasmodic.*

Employment.—*Gravel, catarrh of the urinary bladder, disease of the prostate gland, hæmaturia, rheumatism, gout, dropsy, cutaneous diseases, gonorrhea, glcet, leucorrhea, &c. &c.*

BAROSMIN is a diuretic of the alterative class, and its specific influence is generally more observed in the corrected character of the urine than in its increased flow. The remedial utility of the Barosmin is most especially manifested in the correction of the uric and lithic acid diatheses. Hence its employment is appropriate in all diseases complicated with or taking their rise from a superabundant formation of these acids.

Gravelly affections, characterized by the deposit of a pinkish colored sediment in the urine, offer a wide field for its employment. Both practitioner and patient will sometimes be astonished by the amount of urates eliminated in a few hours under the influence of this remedy. In all cases in which the writer has had occasion to employ preparations of the Buchu, he has found that its efficacy has been materially enhanced by the exhibition of alterative doses of Podophyllin in connection with it.

Catarrh of the bladder is another affection in which I have used this remedy with a very gratifying degree of success. Its peculiar alterative properties are here manifested. It allays the irritation of the mucous surfaces, lessens the amount of mucous voided, and apparently cleanses and heals the abrasions of the mucous surfaces.

In enlargement of the prostate gland, and thickening of the urethral canal, its value as a resolvent can scarcely be estimated. A persevering use of the remedy is requisite in these cases. Among the serious affections to which the urinary apparatus is liable, and in the treatment of which I have employed the preparations of the Buchu with remarkable success, I may mention hæmaturia. The specific tonic property of the remedy is here manifested, and in fact I know of no better tonic remedy for the kidneys under any circumstances.

In rheumatic affections, so frequently dependent upon a uric acid diatheses, I have long employed this remedy with the most satisfactory results. Even in acute rheumatism, after the inflammatory symptoms are measurably subdued, I seldom omit its exhibition.

I have cured many cases of *lumbago* with this remedy, in connection with alterative doses of Podophyllin.

In dropsy it is mainly useful in the asthenic forms, particularly when the kidneys, from want of tone, are tardy in the elimination of the absorbed fluid, or are loaded with uric acid deposits.

In the treatment of cutaneous eruptions, such as salt rheum, eczema, tinea capitis, &c., I consider it a remedy of great value. I have long been of opinion, as heretofore expressed in my writings, that in the treatment of skin diseases diuretics are the best alteratives. I would also mention erysipelas, both acute and chronic, as being a complaint in which I have employed this remedy very successfully.

In the management of gonorrhea, gleet, leucorrhea, and ulcerations of the uterus, this will be found a most valuable auxiliary remedy. I would mention, however, that its employment in the treatment of females will sometimes be attended with a sense of tension and weight in the region of the uterus, and a tendency to prolapsus. When these symptom sappear, the remedy should be suspended.

Administered in warm solution, the Barosmin will generally prove strongly diaphoretic, and the peculiar odor of the plant will be perceptible in the perspiration. It is also very often perceptibie in the urine within an hour or two after being exhibited. The warm infusion will sometimes nauseate.

The dose of the Barosmin is from TWO to FOUR GRAINS. The best vehicle in which to adminster it is water.

## CON. TINC. BAROSMA CRENATA

Derivation and properties similar to the above. I much prefer it on account of its possessing the volatile oil belonging to the plant, and for its convenience of administration and dispensation.

Both these preparations will operate better if a'ministered at least one nour before, or two hours after, meals.

Dose of the Tincture, from TEN TO THIRTY DROPS

# IRISIN.

Derived from *Iris Versicolor*.
Nat. Ord.—*Iridaceæ*.
Sex. Syst.—*Triandria Monogynia*
Common Name.—*Blue Flag*.
Part Used.—The *Root*.
No. of Principles—*four*, viz., *resin, resinoid, alkaloid, and neutral*.
Properties.—*Alterative, resolvent, sialagogue, laxative, diuretic, anti-syphilitic, vermifuge, etc.*
Employment.—*Scrofula, syphilis, gonorrhea, dropsy, rheumatism, glandular swellings, eruptions of the skin, and affections of the liver and spleen.*

IRISIN is justly esteemed as one of our most valuable alteratives. It is eminently resolvent, and exercises a marked influence over the entire glandular system, resolving morbid deposits, quickening the activity of the secreting apparatus, and promoting depuration through the various emunctories. It arouses the functions of the absorbent, venous and lymphatic systems, removes obstructions and corrects aberations of the hepatic and renal functions. As an anti-syphilitic, it has few, if any, superiors. It increases the salivary flow, and has the reputation of producing ptyalism. But a careful distinction must be made between the effects produced by vegetable agencies upon the mucous and salivary glands, and mercurial salivation. The former are nothing more nor less than manifestations of a quickened physiological activity; evidences of special therapeutic stimulus, constituting, oftentimes, a critical conservative effort. No loosening of the teeth, no sponginess of the gums, no putrefactive fetor, no sloughing of the soft parts; increased, but not disordered secretion. On the other hand, mercury induces a pathological condition of the mucous surfaces; provokes a metamorphosis of the vital constituents of the blood and fluids, and favors the formation of vitiated products; altering from good to bad, and from bad to worse; giving rise to congestions, lesions, putrefactive conversions and disorganizations of the organic structures. In the former case we have the evidence of a direct therapeutic stimulus operating upon the vital impressibility of the secreting apparatus, promoting increased activity of its functions for the purpose of eliminating legitimate products. In the latter instance we have an augmented flow of morbid materials resulting from the destructive conversions of the vital constituents by the remedy itself, and which are not the legitimate products of organic metamorphoses. In the former case the remedy itself is the motor-stimulus, while in the latter instance the mercurial corruptions constitute the stimuli of excitement.

We have used the Irisin with good success in the treatment of scrofula. It is peculiarly useful in those cases accompanied with hepatic derangement. The average dose of the Irisin as an alterative is TWO grains, repeated twice or thrice a day. It will generally prove gently laxative in this quantity. In larger doses, say from FOUR to SIX grains, it usually proves cathartic. Its operation is sometimes accompanied with pain and griping, which may be corrected by combining it with stimulants as the Xanthoxylin, Capsicum, ginger, etc. Irisin is mostly employed in combination with other alteratives, as the Stillingin, Corydalin, Phytolacin, Rumin, etc. In many cases it is better to employ it alone and alternate it with tonics. The following formula is of great value in the treatment of hepatic torpor:

℞.
    Irisin........................................ ℈ i.
    Rumin....................................... ℈ ii.

Mix and divide into twenty powders. Dose, one, three times a day. The dose may be increased as occasion requires. The Rumin is a most excellent remedy in chronic disorders of the liver.

Irisin is one of the most excellent remedies we possess for the cure of syphilis. In eradicating the syphiltic virus and correcting the diathesis of the system, it has few equals. Its influences are positive and certain. It may be employed alone and occasionally alternated with other alteratives, or they may be combined as occasion requires. We have prescribed the following formula in many cases, with most excellent results:

℞.
    Irisin........................................
    Phytolacin............................aa. ℈ i.
    Stillingin................................... ℈ ii.

Mix and divide into twenty powders. One of these powders may be given three times daily. We sometimes vary the formula as follows:

℞.
    Irisin ..................................... ℈ ii.
    Corydalin................................. ℈ i.

Mix. From THREE to five grains of this compound may be administered three times per day. Or the following:

℞.
    Irisin.........................................
    Xanthoxylin.......................aa. ʒ ss.

Mix. Dose from TWO to FOUR grains three times a day. In this way we vary the combination to meet the indications of the case. Other of the concentrated medicines, as the Smilacin, Chimaphilin, Alnuin, etc., may be combined with the Irisin to suit particular cases.

Irisin has been found particularly serviceable in the treatment of leucorrhea, congestions of the cervix, ulceration, and other disorders of the uterine system. It is particularly indicated in uterine leucorrhea, in which affection it seems to be of almost specific value. Of course auxiliary treatment must not be neglected.

As an alterative, resolvent, and detergent, the Irisin is highly beneficial in rheumatism, glandular swellings, eruptions of the skin, and in all diseases indicating any peculiar cachexy. We have found it of reliable utility in gonorrhea, gleet, and for the cure of all morbid discharges from the vagina and urethra.

In the treatment of dropsy the Irisin is mainly useful as a resolvent, and for promoting the activity of the absorbent system. In conjunction with the other remedies, it has been successfully employed in the cure of that complaint.

In visceral engorgements and torpor, as of the liver, spleen, etc., the Irisin is a remedy not to be lightly estimated. In doses sufficient to ensure a regular and soluble condition of the bowels, it will be found highly efficacious in chronic hepatic disorders. Also in glandular indurations.

It has been employed in combination with Macrotin, with considerable success for the relief of menstrual suppressions. Two grains of Irisin with HALF a grain of Macrotin will form the average dose, repeated twice or thrice a day.

Irisin is sometimes substituted for Podophyllin when the latter is contraindicated.

# HYDRASTIN.

Derived from *Hydrastis Canadensis.*
Nat. Order.—*Ranunculaceæ.*
Sex. Syst.—*Polyandria Polygamia.*
Common Names.—*Golden Seal, Yellow Puccoon, Ground Raspberry, Tumeric Root, etc.*
Part Used.—The *Root.*
No. of Principles.—*four,* viz., *resin, resinoid, alkaloid,* and *neutral.*
Properties.—*Laxative, chologogue, alterative, resolvent, tonic, diuretic, anti-septic, etc.*
Employment.—*Leucorrhea, gonorrhea, gleet, cystitis, fevers, dyspepsia, constipation, piles, opthalmia, otorrhea, catarrh, and all diseases involving the mucous surfaces.*

HYDRASTIN exercises an especial influence over mucous surfaces. Its action in this respect is so manifest that the indications for its employment cannot be mistaken. Upon the liver it acts with equal certainty and efficacy. As a chologogue and deobstruent it has few equals. In affections of the spleen, mesentery, and abdominal viscera generally, it is an efficient and reliable remedy. Also in scrofula, glandular diseases generally, cutaneous eruptions, indigestion, debility, chronic diarrhea and dysentery, constipation, piles, and all morbid and critical discharges.

Hydrastin has been successfully employed in the cure of leucorrhea. It is of singular efficacy when that complaint is complicated with hepatic aberration. It is employed both internally and externally. The usual dose is from ONE to TWO grains three times a day, increasing the quantity, if more of the laxative effect is needed. For topical use, ONE drachm to ONE pint of boiling water, to be injected tepid or cold, at the option of the patient or practitioner. The same will be found extremely valuable as an injection in gonorrhea, gleet, urethral inflammation, vaginitis, cystitis, hemorrhoids, etc. When considerable inflammation exists, and for injections into the bladder, the infusion should be allowed to stand for a time, in order that the resinoid principle may precipitate, as the neutral and alkaloid principles held in solution by the water are more particularly beneficial in these cases. The resinoid principle possesses a degree of escharotic power, and does not act kindly in certain irritable conditions of the mucous surfaces, proving too stimulating. On the other hand, when the condition is one of coldness and torpor, and when there are exudations of plastic lymph, the action of the resinoid principle is particularly demanded. It is in consequence of this peculiar property of the resinoid principle that Hydrastin is contra-indicated in certain irritated and inflamed conditions of the mucous membranes of the bowels. Its employment under these circumstances will be attended with a troublesome relaxation of the bowels, with griping pains, tenesmus, etc. If employed at all in these cases, it must be accompanied with a plentiful supply of mucilages.

In the treatment of leucorrhea the Hydrastin may be combined with such other remedies as are suited to the indications. We find it valuable joined with Helonin.

℞.
    Hydrastin.

Helonin .............................aa. ℥ i.

Mix. Dose from TWO to FOUR grains three times a day. This combination will be found excellent when indigestion, hepatic torpor, and constipation exist. Stillingin is an invaluable adjunctive when the case has become chronic, and the patient is afflicted with a strumous or scorbutic diathesis.

℞.
    Hydrastin............................. ℥ i.
    Stillingin ............................ ℈ ii.

Mix. Dose, from THREE to FIVE grains three times a day. This treatment should be alternated with an occasional dose of Podophyllin. This formula will be found valuable in gonorrhea, gleet, and catarrh of the bladder.

We have also found the following formula of exceeding utility in leucorrhea when the vaginal secretions were acrid and offensive.

℞.
    Hydrastin............................. ℈ i.
    Super Carb. Soda...................... ℥ i.

Mix. Dose from FOUR to EIGHT grains three times per day. This is one of the best corrective remedies we have ever employed. It is equally advantageous in those forms of indigestion accompanied with acidity, eructations, flatulency, and rectal irritation, and in ulceration of the mucous membranes of the bowels. We value it highly.

The Hydrastin is of inestimable value in the treatment of chronic derangements of the liver and portal circulation. It seems to exercise an especial influence over the portal vein and hepatic structure generally, resolving biliary deposits, removing obstructions, promoting secretion, and giving tone to the various functions. It is eminently chologogue, and may be relied upon with confidence for the relief of hepatic torpor. Its operation is materially enhanced by the administration of an occasional dose of Podophyllin. In some cases they may be combined with advantage.

As a general remedy in the treatment of piles, we know of none better. We have cured many inveterate cases by administering the Hydrastin twice or thrice a day, alternated with an occasional dose of Podophyllin, and using an infusion of the Hydrastin as an injection into the rectum. Perseverance is highly essential to a cure in chronic cases.

Hydrastin has obtained considerable repute as a remedy in intermittent fever. We have employed it to a considerable extent, and in a majority of cases successfully. We have found it most reliable in those cases in which the prolongation of the disease depended upon a disordered condition of the functions of the liver. The administration of a thorough dose of Podophyllin, followed by the judicious use of the Hydrastin, has effected a radical cure in many cases. When a stimulant is required, we combine it with Xanthoxylin, and sometimes with Macrotin. We have used each of the following formulas, and found them all useful:

℞.
    Hydrastin ............................ ℈i.
    Xanthoxylin ......................... ℈ii.

Mix. Dose, from TWO to FOUR grains, once in from two to four hours.

℞.
    Hydrastin ........................ ʒss.
    Macrotin ......................... grs. VIII.

Mix. Dose, from TWO to THREE grains, repeated once in from two to four hours, or as often as the patient can bear. The Macrotin will sometimes produce too much cerebral excitement, and the quantity must be lessened or given at longer intervals. Hydrastin also operates well in conjunction with Cornin.

℞.
    Hydrastin ............................ ℈i.
    Cornin .............................. ℈ii.

Mix. Dose, from THREE to FIVE grains. The anti-periodic power of Hydrastin is feeble, yet it will effect a cure in many diseases characterised by periodicity, by reason of its resolvent, alterative, chologogue, and laxative properties.

In many derangements of the urinary apparatus we have

found the Hydrastin to answer an admirable purpose. In chronic inflammation of the bladder, we deem it one of the most reliable agents of cure. It should be given in small and repeated doses. In congestion of the ureters, chronic suppression of the urine, and gravelly affections, it will be found highly useful. Also in incontinence of the urine, and diabetes. As a tonic in the convalescing stages of fevers, pneumonia, dysentery, and other acute diseases, particularly when a laxative property is needed, the Hydrastin is peculiarly appropriate. It promotes digestion and assimilation, obviates constipation, and gives tone to the depurating functions generally. It has been successfully employed in connection with astringents, as the Geranin, Myricin, Hamamelin, etc., in the treatment of chronic diarrhea and dysentery. Also in ulcerations of the mucous membranes of the stomach and bowels, apthæ, stomatitis, etc. In these cases it operates well in conjunction with Juglandin and Leptandrin.

Externally, the Hydrastin is employed in opthalmia, otorrhea, catarrh, eczema, ulcers, etc. From ONE to TWO drachms may be infused in ONE pint of boiling water, and the resinoid principle allowed to precipitate. It then may be used as a wash in opthalmia, as an injection in otorrhea, and snuffed up the nostrils for the relief of catarrh. We have used it in this way with much benefit. When more of the astringent property is required, it may be joined with Geranin, or Myricin, or Hamamelin, etc., ONE drachm to ONE pint of boiling water. The dry Hydrastin sprinkled upon the surface of an ulcer will act as a mild escharotic, dissolve fungoid growths, and provoke a healthful discharge. We sometimes combine it with Baptisin for this purpose, equal parts. Or with Sanguinarin, Phytolacin, or Trilliin. With Baptisin and Trilliin it forms an excellent application for cancerous and other offensive ulcers, correcting the acrimony and fetor of the discharges. With Baptisin and Trilliin, in infusion, it forms an excellent injection for correcting offensive leucorrheal discharges. Also as an injection into the bowels in diarrhea and dysentery manifesting a tendency to putrescency. Made into an ointment with lard, ONE drachm to the ounce, it is useful in eczema and other cutaneous eruptions, piles, etc. The following we have found excellent for piles, scaly eruptions about the nose, lips, ears, etc.

℞.
    Hydrastin..............................
    Geranin.............................. aa. ʒ ss.
    Gelsemin ..........................grs. XV.
    Lard ................................ ʒj.

Make an ointment. The Hydrastin may be dissolved in alcohol and used with much benefit as a stimulant in obstinate scaly eruptions, opacity of the cornea, enlarged tonsils, syphilitic ulcerations, etc.

We would here add that our experience has demonstrated the Hydrastin to be a valuable remedy in bronchitis, laryngitis, and other affections of the respiratory organs. We give it in doses of from ONE to TWO grains three or four times a day, and use the following gargle:

℞.
    Hydrastin............................. Ʒj.
    Tinc. Myrrh.......................... ʒij.

Mix. One teaspoonful added to a wine glassful of water, and the throat gargled several times a day. A solution of the Hydrastin in water, or its alcoholic tincture diluted in water is also beneficial as a wash in apthous sore mouth, sore throat of scarlatina, etc.

# OIL OF CAPSICUM.

Derived from *Capsicum Annuum.*
Nat. Ord.—*Solanaceæ.*
Sex Syst.—*Pentandria Monogynia.*
Common Name.—*Cayenne Pepper.*
Part Used.—The *Fruit.*
Properties.—*Stimulant, anti-septic, and rubefacient.*
Employment.—*Dyspepsia, constipation, remittent* and *scarlet fevers, coughs, colds, hoarseness, cholera, suspended animation, rheumatism, passive hemorrhages,* and whenever a *pure* and *powerful stimulant* is needed.

THE properties and employment of Capsicum are so well understood that we deem it necessary to say but very little in regard to the oil by which it is represented. This oil is sometimes known by the name of *Capsicin.* It embodies all the properties of the Capsicum, and is employed for the same purposes. It is sometimes joined with other remedies to promote their action, or with Quinine in intermittents, with Podophyllin in cold and indolent conditions of the system, and with other stimulants. It is, perhaps, the very best and purest stimulant known, more prompt but less permanent in its in fluences than many others, as for instance, the Oil of Xanthoxylum. In obstinate hepatic torpor, constipation, paralysis, and in all diseases attended with loss of nervous energy, this remedy is of inestimable utility.

Externally, the Oil of Capsicum is employed as a rubefacient and counter irritant. Except in severe cases of sciatica and neuralgia, the oil is seldom applied pure, but usually dissolved in alcohol.

℞.
    Oil of Capsicum........................ ℨi,
    Alcohol 95 per cent ................... ℨIV,

Mix. This is used as an external application in neuralgia, chillblains, rheumatic pains, chronic sprains, and whenever a powerful stimulating embrocation is needed. Internally, it is administered in doses of from ONE HALF to ONE teaspoonful for pain in the stomach, colic, fainting, etc. Combined with the Oil of Lobelia and dissolved in alcohol, it forms one of the most valuable compounds known in pharmacy. Our formula is as follows:

℞.
    Oil of Capsicum........................
    Oil of Lobelia ........................ aa, ℨi.
    Alcohol 95 per cent.................... ℨIV

Mix. This forms one of the most powerful anti-spasmodic and stimulant preparations known. We use it in locked jaw, apoplexy, convulsions, suspended animation, sun stroke, poisoning, etc., in doses of from ONE teaspoonful to ONE tablespoonful, and repeated at intervals of from twenty minutes to one hour, as occasion requires. In case the patient cannot swallow, it should be administered by injection, and the quantity doubled. As a general thing it may be diluted with water when exhibited, but in trismus and tetanic spasm, and in all cases of difficult deglutition we use the preparation without reduction, by which means we soon overcome the spasm and remove the chief obstacle to further ministrations. In tetanus, when the jaws are set, a small quantity poured between the teeth will, as soon as it reaches the pharynx, relax the spasm and enable the patient to open his mouth, and to swallow. It may also be applied externally to the throat and angles of the jaw, as well as to any part of the system affected with muscular contraction.

We have found this preparation of excellent service in the spasmodic stages of cholera, also in many cases of apoplexy, aided, in the latter instance, by warm stimulating pediluvia, and cold applications to the head. The following plaster applied to the back of the neck, and to the soles of the feet will prove an efficient auxiliary:

℞.
    Oil of Capsicum................gtt. X vel XX.
    Wheat flour....................  ʒ ss.
    Vinegar........................  q. s.

Make a plaster. Spread on paper and apply. We use this plaster in preference to mustard as a counter-irritant. It will not vesicate. Applied to the back of the neck it relieves acute headaches and a tendency to congestion. We have also used it profitably for the relief of pain in various parts of the system, as of the pleura, kidneys, joints, etc., and for the relief of coughs and colds, applied to the chest. Also for the relief of facial neuralgia. When the oil is not at hand we make the plaster as follows:

℞.
    Pul. Capsicum..............  one teaspoonful.
    Wheat flour ................one tablespoonful.
    Vinegar ....................  q. s.

Form a plaster of the proper consistency and spread on paper. The practitioner who becomes fully acquainted with the value of this plaster will seldom use mustard

The following formula will be found excellent for coughs, colds, influenza, hoarseness, sore throat, etc.:

℞.
    Oil of Capsicum..................gtt. V ad X.
    White sugar......................  ʒ ii.

Mucilage of Slippery Elm.......... ℥ IV.

Triturate the oil thoroughly with the sugar and add the mucilage, mixing well together.

Dose, ONE teaspoonful, repeated once in from two to four hours.

The Oil of Capsicum has been found useful in the atonic forms of dyspepsia, both as a radical remedy, and as an adjunct to other medicines. It promotes the flow of the gastric juice, and resolves the viscidity of the secretions. In sluggish conditions of the circulation, plasticity of the blood, venous congestions, etc., it is a remedy of much value. Combined with Trilliin, or other styptics and astringents, it will be found exceedingly valuable for the relief of uterine and other hemorrhages.

The medium dose of the oil is ONE drop, increased or diminished according to the necessities of the case. It enters into the compound Stillingia liniment, as given under the head of **Oil of Stillingia.**

# HAMAMELIN.

Derived from *Hamamelis Virginica.*
Nat. Ord.—*Hamamelaceæ.*
Sex. Syst.—*Pentandria Digynia.*
Common Names.—*Witch Hazel, Winter-bloom,* **Spotted** *Alder,* etc.
Part Used.—The *Bark.*
No. of Principles.—*Two,* viz., *resin* and *neutral.*
Properties.—*Astringent, tonic, and sedative.*
Employment.—*Diarrhea, dysentery,* **hemorrhages, stomatitis,** *leucorrhea, gleet,* etc,

THE Hamamelin is employed with advantage in all cases in which astringents are indicated. It is exceedingly valuable in hemoptysis, hematamesis, hematuria, and in all affections of the mucous surfaces. In diarrhea, dysentery, ulcerations of the stomach and bowels, leucorrhea, gleet, and all excessive mucous discharges, it answers an admirable purpose. Externally, in solution, it is used as a wash in opthalmia, as a gargle in apthous sore mouth, and as an injection in otorrhea, leucorrhea, piles, etc.

The average dose is TWO grains. For injections, etc., from ONE to TWO drachms to ONE pint of boiling water.

It may be combined, when occasion requires, with other astringents, as the Lycopin, Myricin, Geranin, Trilliin, etc., or with tonics, as the Helonin, Fraserin, Cornin, etc. It is very valuable in chronic diarrhea and dysentery, and in mucous hemorrhoids. In injection, it is useful in vaginitis, ulcerations of the cervix, and other affections of the uterine organs.

It has been found beneficial in the latter stages of phthisis for allaying the gastric irritability and restraining the diarrhea. It exercises a peculiar soothing and healing influence over inflamed and abraded mucous surfaces. In solution, it is employed with advantage as a topical application in eczema, tinea capitis, and other cutaneous diseases, bruises, wounds, etc. Made into an ointment with lard it has been used for the same purposes; also in piles.

The solution will be found useful as an injection in prolapsus of the womb, rectum, etc.

The doses may be increased to FIVE and even to TEN grains, in several cases, with safety and advantage.

# EUPHORBIN.

Derived from *Euphorbia Corollata.*
Nat. Ord.—*Euphorbiaceæ.*
Sex. Syst.—*Dodecandria Trigynia.*
Common Names.—*Bowman's Root, Blooming Spurge, etc.*
Part Used.—*The Root.*
No. of Principles.—*Two,* viz., *resinoid* and *neutral.*
Properties.—*Emetic, cathartic, diaphoretic, expectorant, and vermifuge.*
Employment.—*Fevers, dropsy, diarrhea, dysentery, biliary congestions, worms, etc.*

THE Euphorbia is a reliable acquisition to our indigenous materia medica, and fulfills many important indications. In small, repeated doses, it acts as a diaphoretic, inducing free perspiration, deterges the mucous coats of the stomach and bowels, stimulates the functions of the liver, and corrects the tendency to colliquitive diarrheal discharges. In large doses it is emetic and cathartic. If an undue amount of acidity predominates in the stomach, its emetic powers are suspended, and it passes off by the bowels. It is for this reason that, when administered as an emetic, it has obtained the reputation of being uncertain in its operation. The necessity of neutralising undue acidity previous to its administration will, therefore, be apparent. When administered as an emetic, it will generally vomit without exciting any previous nausea while, at other times, considerable prostration of the muscular system with lingering nausea will be observed, paleness of the countenance, and a cool, moist state of the skin, from which, however, the patient rapidly recovers as soon as the medicine has operated upon the bowels. Its action in this respect may generally be corrected by the administration of alkalies, or of a quick cathartic, as the Jalapin. We deem the Euphorbin one of the most powerful, and, at the same time, safest revulsive remedies that can be administered for the relief of cerebral congestions. It excites, powerfully, the absorbent and venous systems, and is, therefore, frequently employed for the removal of dropsical effusions, removing them when other means fail. Combined with Podophyllin, as given under that head, its efficacy is enhanced, and, so combined, is employed in the forming stages of typhoid and other fevers, dropsy, cerebral congestions, obstinate menstrual suppressions, and for the removal of biliary concretions.

We have found the Euphorbin of much utility in the treatment of cholera infantum, diarrhea and dysentery. It seems to exercise a peculiar control over the glandular structure of the intestinal canal, correcting and giving tone to the action of the secreting vessels, and promoting assimilation of the fecal matters. We have administered it in cases of cholera infantum when the alvine discharges were watery, copious, and offensive, and had, as the result of its operation, well assimilated stools, without fetor. It seems to possess considerable power in correcting a tendency to putrescency. We have been unable to discover that the Euphorbin acts as a special irritant upon the bowels, but, on the contrary, esteem it as a corrective of irritation. Our observations of its operation have led us to the conclusion that the irritation sometimes observable is the result of an increased activity on the part of the eliminating vessels of the alimentary canal, and the consequent depuration of certain morbid and acrid materials from the blood, which, being brought in contact with the mucous surfaces, constitute an extraneous cause of excitement. It may be, also, when the root has been administered in substance, that the non-medicinal constituents have undergone a fermentative decomposition, and given rise to products that operated as special irritants. Or the ligneous portions may have operated mechanically. At any rate, we have found the Euphorbin to control rather than to excite irritation.

The average dose of the Euphorbin, as a diaphoretic, is from ONE FOURTH to ONE grain, repeated at intervals of from one to three hours. As an emetic and cathartic, from TWO to THREE grains. It may be combined with Asclepin to increase its diaphoretic and expectorant powers.

℞.
    Euphorbin.......................... grs. X.
    Asclepin ............................grs. XL.

Mix. From ONE to TWO grains of this combination may be administered once in from one to three hours, and will be found exceedingly valuable in fevers, acute rheumatism, pneumonia, pleuritis, acute bronchitis, dysentery, etc. If nausea arise, diminish the dose, or exhibit at longer intervals. In many cases it is desirable to provoke and maintain a degree of nausea, for which purpose nothing better can be devised.

In the treatment of dropsy, the Euphorbin is usually administered in full doses, say from TWO to FOUR grains. Its employment, however, will be contra-indicated in cases accompanied with much debility. It may be combined with Podophyllin or Jalapin, at the option of the practitioner.

For the removal of worms, we usually give from ONE HALF to ONE grain twice or three times a day, or sufficiently often to keep the bowels somewhat relaxed, and continue it for three or four days at a time. Even when no worms are removed, its administration results in decided benefit to the patient. We value it exceedingly in the treatment of the indigestion of children, and for the removal of all that train of symptoms which is usually supposed to indicate the existence of worms These are, loss of appetite, or it may be variable, voracious at times, and none at others, furred tongue, feverishness, fetid breath, bloating of the stomach, constipation, or, on the contrary, a troublesome diarrhea, emaciation, peevishness, wakefulness or disturbed sleep, etc. For the relief of these symptoms we rely with much confidence upon the judicious employment of the Euphorbin. As a general thing no other medicine will be needed, but, when indicated, tonics may be employed in connection.

We can conscientiously recommend the Euphorbin to the profession as a remedy entitled to their confidence

# LYCOPIN.

Derived from *Lycopus Virginicus.*
Nat. Order.—*Laminaceœ.*
Sex. Syst.—*Diandria Monogynia.*
Common Names.—*Water Horehound, Bugle Weed, Sweet Bugle,* etc.
Part Used.—The *Herb*
No. of Principles.—*two,* viz., *resinoid* and *neutral.*
Properties.—*Astringent, styptic, sedative* and *tonic.*
Employment.—*Incipient phthisis, hemoptysis, hematamesis, hematuria, uterine* and other *hemorrhages, diabetes, chronic diarrhea* and *dysentery, cardiac affections,* etc.

The Lycopin is, with us, an exceedingly valuable remedy. Its action is peculiar and positive. It exercises a special influence over the respiratory, cardiac, and renal functions, and obviates a tendency to sanguineous exudations and effusions. No agent yet discovered can compare with it in efficacy as a radical remedy in the treatment of hemorrhage of the lungs. In this complaint it seems to be almost a specific. We have used the plant and its preparations long and successfully, and can speak with authority. It is an arterial sedative of the most valuable kind, reducing the force and frequency of the pulse when abnormally excited, and its operation is unattended with any symptoms of narcotism. It resolves congestions of the capillary and venous plexuses, and invigorates and gives tone to the capillary structure generally. It is a tonic of more than ordinary efficiency, invigorating the appetite, promoting digestion and assimilation, and allaying gastric and enteric irritability. It cleanses and heals abrasions and ulcerations of the mucous surfaces, and gives tone to the muscular fibres. Upon the skin and kidneys it operates in a peculiar and desirable manner, restoring the secreting power, and harmonising and giving tone to those functions.

Lycopin is the most reliable remedy for the radical cure of hemoptysis that we have ever employed. We give it in TWO grain doses three or four times a day, preferring to administer it in water. In severe hemorrhages we administer it every thirty or sixty minutes until relief is afforded, and then continue as above stated. The same directions will apply in hemorrhages of every kind. The doses may be increased, or repeated at shorter intervals when the urgency of the symptoms render it necessary. It may be combined, if desired, with other styptics and astringents, as the oil of Erigeron, Trilliin, Geranin, etc., but we have generally found the Lycopin competent without the aid of auxiliaries. Lycopin has been found serviceable in incipient phthisis, abating the febrile tendency, promoting expectoration, strengthening digestion, aiding cutaneous and renal depuration, and restraining a tendency to hemorrhage. It is employed in the manner directed above.

For the cure of ulcerations of the stomach and bowels, chronic diarrhea and dysentery, and diseases of the mucous surfaces generally, the Lycopin should be given in doses of TWO or THREE grains three times a day, and alternated with Leptandrin or Juglandin in sufficient doses to ensure a soluble condition of the bowels. It may be combined, when in the judgment of the practitioner it is advisable, with Myricin, or Rhusin, or

Geranin, and other astringents, or with tonics, as the Fraserin, Cornin, etc.

Lycopin has been found of remarkable efficacy in diabetes. We give it in doses of from TWO to FOUR grains three times a day, and regulate the bowels with Hydrastin. A suitable diet and regimen must be adhered to. The alkaline sponge bath must be employed two or three times a week, and the food, for a time, consist mostly of animal gelatine. An occasional alterative dose of Podophyllin and Leptandrin will much facilitate the operation of the Lycopin.

The properties above ascribed to the Lycopin are positive and uniform, and the remedy may be relied upon to accomplish all we have said for it. Its value once known to the practitioner, he will consider his therapeutic repertory incomplete without it. Its operation is promoted by the administration of warm diluent infusions.

# FRASERIN.

**Derived** from *Frasera Carolinensis.*
Nat. Ord.—*Gentianaceæ.*
Sex. Syst.—*Pentandria Monogynia.*
Common Name.—*American Colombo.*
Part Used.—The *Root.*
No. of Principles—*three*, viz., *resin, neutral* and *mucresin.*
Properties.—*Tonic, stimulant,* and *mildly astringent*
Employment.—*Indigestion, debility, diarrhea, night-sweats, hysteria, gravelly disorders, etc.*

FRASERIN is a special tonic and stimulant to the digestive organs, and particularly to the mucous membranes of the liver and other viscera concerned in digestion and assimilation. It possesses no laxative properties, but, on the contrary, is slightly astringent. As a tonic it will be accepted by the stomach when other tonics are rejected, and its employment is admissible in the most extreme cases of debility, by virtue of its kindly influences upon this and the surrounding organs.

We employ Fraserin in atony of the digestive organs, and in all cases of disordered secretion manifested in and by that apparatus. It is of exceeding utility in the convalescing stages of fevers, diarrhea, dysentery and cholera infantum, and in all cases in which the system has been exhausted by profuse colliquitive discharges. In all cases of viceral debility, whether primary or induced by copious and exhaustive secretion, Fraserin is equally appropriate. It is also useful in all cases in which the secretions evince a septic tendency, having considerable power as an antiseptic.

Fraserin is of great service in the treatment of bilious diseases occurring in hot climates. In the latter stages of bilious and asthenic dysenteries, and even in cholera, we can recommend the Fraserin as entitled to much confidence. Also in jaundice accompanied with extreme debility, mucous hemorrhoids, dyspepsia, etc. Hypochondriacal and hysterical affections are also relieved by it. Colliquitive diarrheas are frequently cured with Fraserin alone. In arthritic and gravelly affections accompanied with debility of the digestive organs, the Fraserin will be found an excellent remedy.

The average dose of the Fraserin is FOUR grains, but will vary from TWO to TEN. It is best administered dissolved in warm water. It may be combined with **aromatics** and with **anti-spasmodics**, as the Dioscorein.

# XANTHOXYLIN.

Derived from *Xanthoxylum Fraxineum.*
Nat. Ord.—*Xanthoxylaceæ.*
Sex Syst.—*Diœcia Pentandria.*
Common Name.—*Prickly Ash.*
Part Used.—The *Bark.*
No. of Principles—*two,* viz., *resinoid* and *neutral.*
Properties.—*Stimulant, tonic, alterative* and *sialagogue.*
Employment.—*Rheumatism, scrofula, paralysis, indigestion, colic, syphilis, etc.*

BESIDES the two active principles above named, the bark of the Xanthoxylum yields an oil, which will be treated of next in order.

The Xanthoxylin possesses the properties enumerated above in an eminent degree, and will be found highly useful in the diseases mentioned. We have used it extensively, and esteem it a remedy of great value. It is a stimulant of the most permanent kind, having considerable control over the circulation, which it quickens and maintains. It also gives activity to the muscular fibres of the stomach and bowels, pro motes the flow of the saliva, gastric, and other digestive juices, and restores the proper secreting power of the mucous surfaces

Xanthoxylin is a remedy of great value in the treatment of the atonic form of indigestion, scrofula, chronic rheumatism, paralysis, general debility, cutaneous eruptions, ulcers, chronic diarrhea, dysentery, ulcerations of the stomach and bowels, syphilis, gleet, leucorrhea, etc., and for the correction of all languid conditions of the system. It enhances the efficacy and gives permanency to the influences of other stimulants and tonics.

The average dose of the Xanthoxylin is from TWO to FOUR grains. It may be combined with other remedies when indicated, or alternated with suitable agents, at the option of the practitioner It operates well in combination with Stillingin in syphilis, chronic diarrhea, gleet, etc. With Macrotin we have found it highly beneficial in chronic rheumatism. In combination with Fraserin it will be found highly serviceable in the convalescing stages of dysentery, cholera infantum, and other bowel disorders. Other combinations are pointed out in the course of this work, and need not be repeated here.

# OIL OF XANTHOXYLUM.

Derivation same as above.

This oil possesses properties analogous to the above, being, however, more decidedly stimulating, with less of the alterative and tonic proporties. Its use is more appropriate in asthenic than in sthenic conditions, as it is apt to produce too much irritation of the mucous surfaces. It is employed in colic, chronic rheumatism, syphilis, etc. It may be combined with Irisin and Phytolacin and formed into pills for the treatment of the diseases last mentioned.

The average dose of the oil is from TWO to FIVE drops. It may be dissolved in alcohol and so incorporated with other mixtures when desired, or taken upon sugar, or suspended in mucilage.

# CON. TINC. XANTHOXYLUM FRAX.

In this preparation we have embodied the entire therapeutic value of the bark, and which may be used for all the purposes of the crude article. It is positive and uniform in strength, and convenient of administration. The average dose is from TWO to FOUR drops. It is more active than the Xanthoxylin, but not so appropriate in the treatment of infantile disorders, nor in cases of great debility. The Xanthoxylin, being deprived of the oil, is easily soluble and readily assimilated, hence more compatible in enfeebled conditions, as the beneficial effects of remedies depend somewhat upon the ability of the system to appropriate them. We sometimes employ the Con. Tinc. in combination with Leptandrin, Populin, Juglandin, etc., as noticed under those heads.

# SANGUINARIN.

Derived from *Sanguinaria Canadensis*
Nat. Ord.—*Papaveraceæ.*
Sex. Syst.—*Polyandria Monogynia.*
Common Names.—*Blood Root, Red Puccoon,* etc.
Part Used.—*The Root.*
No. of Principles, *four*, viz., *resin, resinoid, alkaloid* and *neutral.*

Properties.—*Emetic, sedative, febrifuge, stimulant, tonic, alterative, resolvent, diuretic, emmenagogue, detergent, antiseptic, expectorant, laxative, errhine,* and *escharotic.*

Employment.—*Fevers, pneumonia, croup, influenza, rheumatism, amenorrhea, hooping cough, asthma, constipation, gravel, scrofula, jaundice, dropsy, dyspepsia,* etc.

VARIOUS preparations of the Sanguinarin have been before the profession, each claiming to represent the medicinal properties of the plant, but, being composed of single isolated principles, they failed to do so. We have had what was called the *alkaloid* principle under the name of *Sanguinarina*, the *alka-resinoid* principle under the title of *Sanguinarin*, etc., but each were fractional and indefinite preparations entitled to no confidence whatever. The Sanguinaria Canadensis is truly a valuable plant, highly esteemed by the profession, and one of which a concentrated equivalent is highly desirable. In the Sanguinarin now under consideration, we believe this *desideratum* to have been accomplished. The four active proximate principles of which it is composed embody the entire therapeutic constitution of the plant, and in their physiological influences demonstrate the fact of their equivalency.

In small, continued doses, the Sanguinarin is a stimulating diaphoretic, resolvent, alterative, and diuretic. Under its immediate influence the pulse rises, but subsequently sinks somewhat below the normal standard, for which reason the Sanguinarin has acquired the reputation of being narcotic. We are inclined to view the depression of the circulation as a secondary influence, resulting from the relief of certain abnormal conditions upon which arterial excitement was dependent, such as plasticity of the blood, retention of effete matters, capillary congestion, etc., and which have been obviated by the resolvent, diaphoretic, and other properties of the Sanguinarin. Cutaneous depuration is powerfully promoted by the Sanguinarin, hence it is of great value in all cases in which such a property is required, as in fevers, rheumatism, skin diseases, etc. The Sanguinarin ranks high as an expectorant, for which purpose it should be given in small and frequently repeated doses. Few remedies exercise a more decided influence upon the urinary apparatus, upon which it displays its peculiar powers as an alterative. In obstinate gravelly affections, and in functional inactivity of the kidneys it is peculiarly serviceable. It is equally efficient in promoting the secretions of the serous as well as of the mucous membranes, hence is a valuable remedy in the treatment of chronic pleuritis, peritonitis, and other affections of the serous surfaces. Over the capillary circulation it exercises a wonderful control, operating as a vascular excitant, and in cold and languid conditions of the circulation, manifested by coldness of the extremities, a relaxed and pallid appearance of the skin, great sensitiveness to atmospheric changes, etc., it will be found one of the most reliable remedies possible to employ.

Sanguinarin resolves the plasticity of the venous blood, and stimulates the venous, absorbent and lymphatic vessels and glands. It is, for these reasons, a valuable remedy in the treatment of dropsy, particularly the asthenic forms, arousing the system from its torpor, and invigorating the functions of secretion and depuration. The liver comes within the especial province of the sanative influences of the Sanguinarin, and in all cases of hepatic torpor, jaundice, biliary concretions, chronic hepatitis, and other abnormal conditions of that organ, the practitioner will find it a remedy worthy of his highest confidence.

The emmenagogue properties of the Sanguinarin are marked and decided, and in chronic amenorrhea have proved of exceeding utility. In all atonic conditions of the uterus and its appendages the Sanguinarin will be found an efficient auxiliary. It is decidedly anti-septic, and is beneficially employed in offensive leucorrheal discharges, ulcerations of the cervix, chancres, buboes, etc.

In larger doses the Sanguinarin operates as a prompt and efficient emetic, and is employed in croup, pneumonia, fevers, to eject poisons, and whenever prompt emesis is desirable. Its operation as an emetic is sometimes attended with a severe burning sensation and pain in the stomach, which lasts for a considerable time after the medicine has operated. This effect may be obviated in a measure by the abundant use of mucilages. The Sanguinarin possesses a considerable degree of escharotic power, hence its use is contra-indicated in gastritis and enteritis, and whenever we have occasion to suspect abrasion or ulceration of the mucous surfaces of the stomach or bowels. When used as an emetic it should be thoroughly triturated with Eupatorin Perfo., and diffused in plenty of warm water, or a thin gruel of corn meal. It may sometimes be usefully combined with the Wine Tinc. of Lobelia, particularly in croup, asthma and pneumonia. It has a tendency to quicken the operation of other emetics.

Of the special employment of the Sanguinarin we note as follows. In all fevers denoting a languid condition of the vital forces its employment is peculiarly appropriate. As a diaphoretic it is scarcely excelled. It belongs to the class of nauseants, hence its administration must be governed accordingly. For favoring the development of the exanthema in eruptive fevers, we know of nothing better. We have used it with marked success in the treatment of scarlatina. The average dose is from ONE EIGHTH to ONE FOURTH of ONE grain, repeated once in one or two hours as occasion requires. If nausea arise, and it be not desirable, the doses may be diminished, or administered less frequently. In many cases a degree of nausea is necessary to the overcoming of capillary constriction, in which event the Sanguinarin will be found to answer an admirable purpose. It will operate more efficiently as a diaphoretic and febrifuge if administered in warm water. Joined with Asclepin, its efficacy will be materially enhanced. We observe the following formula:

℞.
  Sanguinarin.............................grs. ii.
  Asclepin............................... ʒ ss.
  Warm water........................... ℥ iv.

Triturate the Sanguinarin thoroughly with the Asclepin nd add the water. Dose, one teaspoonful every hour. The doses and frequency of repetition are to be governed by the necessities of the case. It is desirable to excite and maintain a gentle and permanent diaphoresis. This preparation may also be employed with great advantage in pneumonia, influenza, bronchitis, asthma, whooping cough, and other affections of the respiratory organs. The expectorant power of the San guinarin is considerable, and is particularly displayed when the pulmonary secretions are viscid from retention. In incipient phthisis, asthma, influenza, bronchitis and other affections of the respiratory apparatus, the Sanguinarin may be given in doses of from ONE EIGHTH to ONE HALF of ONE GRAIN three or four times daily. Suitable combinations may be effected when existing symptoms indicate their necessity.

Thus in asthma, the convalescing stages of croup, influenza, etc., the Sanguinarin will act exceedingly well in combination with Eupatorin Purpu. We observe the following proportions:

℞.
    Sanguinarin........................grs. ii.
    Eupatorin Purpu..................... ℨ ss.

Mix and divide into SIXTEEN powders. One of these may be given in from two to four hours. Valuable in hooping cough, and in all cases of dyspnea. If these powders be alternated with suitable doses of Asclepin, their efficacy will be much enhanced. The latter will assist in promoting the action of the cutaneous exhalents. When tonics are indicated, the Sanguinarin may be combined with Prunin, or Fraserin, or Cornin, etc.

Sanguinarin is efficient in overcoming hepatic torpor, in which affection it may be given in doses of from ONE EIGHTH to ONE grain twice a day. Joined with Podophyllin, or Leptandrin, or Phytolacin, etc., it will promote their action, and so combined may be employed in chronic and obstinate cases of constipation, visceral enlargements, jaundice, gravel, and in all cases requiring a powerful alterative, resolvent, and deobstruent remedy.

In the treatment of secondary and tertiary syphilis, the Sanguinarin has been found of great service. In all cold and languid conditions of the system it is useful for arousing the impressibility of the nerves, and so preparing the way for other remedies. In the above mentioned disease it may be combined with other alteratives, as the Stillingin, Corydalin, Phytolacin, Irisin, etc. In eczema, herpes, syphilitic eruptions, and other diseases of the skin, it will be found to operate admirably in connection with Cerasein. The Sanguinarin may be given in doses of from ONE FOURTH to ONE HALF grain twice a day, and alternated with FIVE grain doses of Cerasein.

As an emmenagogue, the Sanguinarin has acquired considerable repute. In cases of debility it should be used in connection with suitable tonics, as the Fraserin, Cornin, Iron, etc. The following formula constitutes the most powerful emmenagogue remedy with which we are acquainted:

℞.
    Sanguinarin .......................... grs. ii.
    Macrotin ............................. grs. viii.
    Baptisin ............................. grs. xvi.

Mix and divide into sixteen powders. One of these may be given morning and evening. In simple amenorrhea, not accompanied with debility or other complications, this remedy will be found one of the most efficient that can be employed. The exhibition of an occasional dose of Podophyllin will render success almost certain. The Sanguinarin, as with all other forcing remedies, is contra-indicated in anemic habits.

We might specify many other forms of disease in which the Sanguinarin may be beneficially employed, but we are aware that the profession are already quite well acquainted with the virtues of the plant, and, as the Sanguinarin is its true concentrated equivalent, they have but to transfer that knowledge to the preparation under consideration. Those who are not familiar with its properties and employment may, by attentively studying the history we have given of its dynamic influences, easily comprehend its adaptation.

Externally, the Sanguinarin is beneficially employed for a variety of purposes. It possesses considerable escharotic power, and is also anti-septic. It is applied to nasal and uterine polypi, and in some cases will disorganise them. Applied to the surface of foul and indolent ulcers, it cleanses and disposes them to heal. It may be combined with Hydrastin, Baptisin, Trilliin, or Phytolacin. In solution, in water, from TEN to FORTY grains to the pint, there is, perhaps, nothing better as a gargle in the sore throat of scarlatina. Also in other ulcerative affections of the mouth and throat. In scaly eruptions of the skin, dissolved in alcohol or strong vinegar, it has been employed with much success. Also, in combination with caustics, in the treatment of cancers and malignant ulcers.

# PRUNIN.

*Derived* from *Prunus Virginiana.* (*Cerasus Serotina*)
Nat. Ord.—*Drupaceæ.*
Sex. Syst.—*Icosandria Monogynia.*
Common Names.—*Wild Cherry, Black Cherry,* etc.
Part Used.—The *Bark.*
No. of Principles.—*Three,* viz., *resinoid, neutral,* and *amygdalin.*
Properties.—*Stimulant, tonic, expectorant, and, in large doses, sedative.*
Employment.—*Coughs, colds, incipient phthisis, dyspepsia, hectic fever, debility, scrofula,* etc.

MUCH uncertainty has hitherto attended the question, in what peculiar principle resides the active properties of the wild cherry bark?" Some have supposed that its medicinal value depended upon the presence of hydrocyanic acid, viewed by early writers as an educt, but latterly, and correctly, as a product of the decomposition of amygdalin. Others have attributed its medicinal influences to a portion of the amygdalin remaining undecomposed. Various conjectures in regard to its active constituents have prevailed at times, all open to objections, and all lacking confirmation, until, at last, the more philosophical conclusion was, that the more valuable therapeutic properties resided in some "yet undiscovered principle." Such was truly the case. We now have the pleasure of presenting the profession with that "undiscovered principle" in the *neutral* proximate active constituent of the Prunin under consideration. In the neutral resides the chief tonic power of the bark. It is perfectly soluble, and is the principle yielded to infusions and decoctions. But it has been observed that decoctions of the bark seem deficient in medicinal value. By referring to the article on infusions and decoctions, in the first part of this work, the reader will there find the cause explained, namely, the conversion of the neutral principle into apotheme. But our space will not permit our going into a fuller elucidation of the subject, and we shall rely upon the therapeutic integrity of the Prunin to sustain the statements we have put forth.

Prunin is a valuable stimulant, tonic, and expectorant, when given in small and repeated doses, and an arterial sedative of considerable efficacy when given in larger doses. Its special tonic influences seem to be directed mainly to the digestive and assimilative apparatus, promoting activity and giving vigor in the performance of their functions. Hence it is valuable in cases of enfeebled digestion, particularly in the convalescing stages of pneumonia, fevers, and other acute diseases, incipient phthisis, and in all cases in which the additional property of an expectorant is indicated. In the asthenic forms of dyspepsia it has been found peculiarly serviceable. In hectic fever it has likewise been employed with much benefit. It seems to give tone to the cutaneous capillary structure, and to restrain the tendency to colliquitive sweats. It promotes the appetite, strengthens digestion, calms the irritability of the nervous system, and allays inordinate action of the heart and arterial vessels. From these considerations of its dynamic influences, its range of application may be easily deduced. The average dose of the Prunin, as a tonic, is TWO grains. As a sedative from FOUR to EIGHT. The frequency of the repetition must be governed by the judgment of the practitioner. As an expectorant, we give from ONE to TWO grains every two hours.

Prunin admits of many appropriate combinations, which may be used to advantage in the treatment of complicated cases. Thus, when we wish a diaphoretic, expectorant, and tonic influence, we combine it with Asclepin.

℞.
    Prunin.................................... ℨj.
    Asclepin.................................. ℨss.

Mix. From TWO to THREE grains of this compound may be exhibited once in two or three hours, as may be necessary. It will be found valuable in the convalescing stages of pneumonia, bronchitis, influenza, and in phthisis when the cough is dry and expectoration difficult. Also in hooping cough, chronic cough, and some forms of asthma, as well as in the asthenic stages of croup.

Prunin may also be advantageously combined with Eupatorin Purpu., forming a useful remedy in the treatment of dropsical affections. We employ it as follows:

℞.
    Prunin.....................................
    Eupatorin Purpu...................... aa. ℨss.

Mix. From TWO to FIVE grains of this mixture may be given once in six hours. Valuable in gravely disorders, catarrh of the bladder, leucorrhea, and atony of the urinary and generative apparatus. We sometimes employ it combined with Senecin.

℞.
    Prunin.....................................
    Senecin ............................... aa. ℨj.

Mix. Dose, from TWO to FIVE grains three times a day. In cases of amenorrhea, dysmenorrhea, leucorrhea, and other uterine disorders accompanied with feeble digestion, this remedy is of exceeding utility.

It will be sometimes observed that the employment of the

Prunin in affections of the respiratory system will prove objectionable on account of its producing constriction of the chest and difficult respiration. This effect seems to arise from some constitutional peculiarity of the patient, the Prunin proving too much of a stimulant. If other indications are had for its employment, this influence may be obviated in a measure by combining it with anti-spasmodics and expectorants, as the Asclepin, Eupatorin Purpu., Veratrin, Cypripedin, etc.

Prunin may be joined with other tonics with advantage in particular cases, as with Fraserin in the convalescing stages of diarrhea, dysentery, and cholera infantum; and with Hydrastin or Euonymin when a laxative property is needed. It is best administered in water.

# MENISPERMIN.

Derived from *Menispermum Canadense.*
Nat. Ord.—*Menispermaceœ.*
Sex. Syst.—*Diœcia Polyandria.*
Common Names.—*Yellow Parilla, Moonseed, etc.*
Part Used.—The *Root.*
No. of Principles—*three,* viz., *resinoid, alkaloid,* and *neutral.*

Properties.—*Alterative, tonic, laxative, diuretic,* and *stimulant.*

Employment.—*Scrofula, syphilitic infections, cutaneous eruptions, gout, rheumatism, hepatic torpor, constipation, loss of appetite, indigestion, glandular enlargements, etc.*

THE Menispermin is a remedy of positive and remarkable value. We have employed it with a great degree of satisfaction in the treatment of a variety of affections. As an alterative and resolvent, it deserves to be ranked with the best in the materia medica. It excites the action of the glandular system in a peculiar manner, resolving vitiated deposits, correcting the action of the secretory functions, stimulates the venous, absorbent, and lymphatic vessels, and promotes depuration through the various channels. It is an alterative diuretic of well attested efficacy, and a laxative of more than ordinary value, operating without irritation. Upon the functions of the skin it seems to exercise an especial influence, promoting cutaneous depuration in a peculiar manner. At the same time it imparts a peculiar toning influence to all parts of the organism involved in its therapeutic control. It is especially useful in atonic conditions of the system, as it seems to possess the power of promoting its own appropriation. It stimulates the entire vascular system, and increases the force and frequency of the pulse. In very large doses, it proves emetic and cathartic.

Among the diseases in which the Menispermin has been found valuable, we would mention scrofula. From a consideration of the foregoing enumeration of its physiological influences, its appropriateness in the treatment of strumous diseases will be manifest. It increases the appetite, strengthens digestion, promotes absorption and assimilation, resolves viscid deposits, and imparts activity and tone to the entire depurative structure of the system. The medium dose of the Menispermin, in these cases, is TWO grains, increased to FIVE if more of the laxative property is needed, and repeated twice or three times a day. When the indications render it admissible, it may be joined with other alteratives, as the Stillingin, Irisin, Ampelopsin, etc. As a general thing, however, we prefer to alternate it with such other remedies as may be appropriate in the case. In strumous affections, complicated with suppression of the menses, it operates well in connection with Helonin and Senecin. In the treatment of chlorosis, it should be joined with Iron. In the treatment of the asthenic forms of scrofula, we deem the Menispermin one of our most valuable agents. It is a stimulating alterative and tonic of a high order of therapeutic value, and peculiarly appropriate in all atonic conditions of the venous, lymphatic, and glandular systems.

Menispermin has been found of marked utility in the cure

of syphilis, particularly for the relief of that peculiar train of symptoms termed mercurio-syphilitic. It may be joined with Stillingin, Phytolacin, Irisin, Corydalin, Ampelopsin, etc., at the option of the practitioner, and alternated with suitable doses of Podophyllin.

In the treatment of cutaneous diseases, we value the Menispermin highly. Its action upon the skin is remarkable and peculiar, restoring the functional activity and integrity of the entire cutaneous structure. When indicated, it should be combined with Iron, not omitting the alkaline sponge bath. Alternated with the tincture of the chloride of Iron, it will be found highly efficacious in scaly eruptions of the skin, herpes, erysipelas, etc.

We have employed the Menispermin with much success in the atonic forms of dyspepsia, and in those cases of enfeebled digestion following attacks of acute diseases. Particularly when constipation, loss of appetite, and a feeble circulation are present, will it prove of peculiar utility. It may be joined with Hydrastin, thus forming one of the best combinations with which we are acquainted for fulfilling the indications above mentioned.

On account of the stimulant, tonic, alterative and resolvent properties of the Menispermin, it is highly beneficial in the treatment of chronic rheumatism. When desirable to increase the stimulant effect, it may be joined with Xanthoxylin; and when the circulation is much enfeebled, accompanied with coldness of the extremities, with Sanguinarin. Menispermin is highly useful in gravelly disorders and dropsy. It stimulates the functions of the absorbent system, and promotes the depurative action of the kidneys, resolving calculous deposits, and favoring their expulsion. In all affections of the glandular system we would recommend it **as worthy of the confidence of the profession.**

**Medium dose, TWO grains.**

# OIL OF SOLIDAGO.

---

Derived from *Solidago Odora.*
Nat. Order.—*Asteraceæ.*
Sex. Syst.—*Syngenesia Superflua.*
Common Name.—*Sweet Scented Goldenrod.*
Part Used.—The *Leaves.*
Properties.—*Aromatic, stimulant, carminative,* and *diuretic.*

Employment.—*Pain in the stomach and bowels, flatulence, suppression of urine, inflammation of the kidneys and bladder, and for inhalation in diseases of the respiratory organs.*

THE oil of Solidago is a mild but efficient remedy in the complaints above mentioned. It may be given in doses of from TWO to FIVE drops, and repeated every thirty or sixty minutes until relief is obtained. It is peculiarly appropriate in the treatment of the colicky pains of infants, being mild and unirritating in its operation. It is likewise highly beneficial in the treatment of suppression of the urine occurring in children and infants. It is better, as a general thing, to dissolve the oil in alcohol for employment in these cases.

℞.
    Oil of Solidago   •   •       -   ʒ i.
    Alcohol         -       -   ʒ viii.

From FIVE to TWENTY drops, or more, may be given at a dose, and repeated at suitable intervals. The same will be found excellent for flatulent pain in the stomach and bowels, faintness, etc., in adults. The dose is from ONE-HALF to ONE teaspoonful. Equal parts of the above ticture, Holland Gin, and Swt. Spts. Nitre, mixed and given in doses of from a teaspoonful to a tablespoonful, will be found highly efficacious for the relief of suppression and retention of urine, and inflammation of the kidneys and bladder. In the latter affections it should be accompanied with a plentiful supply of mucilages. We have employed the Oil of Solidago for the purposes of inhalation in the affections of the respiratory organs, and with much benefit. The oil should first be dissolved in alcohol, in the proportion above directed. One teaspoonful of this tincture may be employed for each inhalation. It relaxes constriction of the lungs, soothes the pulmonary surfaces, and promotes expectoration. It is useful in bronchitis, asthma, influenza, catarrh, pneumonia, and phthisis. The inhalation may be repeated four or five times daily.

When the alcohol is objectionable, the oil may be taken on sugar or suspended in mucilage.

# SMILACIN.

Derived from *Smilax Officinalis.*
Nat. Ord.—*Smilaceæ.*
Sex Syst.—*Diœcia Hexandria*
Common Name.—*Sarsaparilla.*
Part Used.—*The Root.*
No. of Principles.—*two,* viz., *resinoid* and *neutral.*
Properties.—*Alterative, resolvent,* and *detergent.*
Employment.—*Scrofula, venerial diseases, rheumatism, cutaneous diseases etc.*

NOTWITHSTANDING the Sarsaparilla is a remedial agent of variable reputation, it really possesses most valuable properties as an alterative and restorative. Many facts can be adduced of the want of therapeutic uniformity manifested in the history of this plant, to a few of which we would wish briefly to call attention. In the first place, the reader will please to call to mind the history we have given of the neutral proximate active principles. The variable amount of this constituent, whether it be owing to the fact of its imperfect development at the time the plant was collected, or to chemical reactions afterwards transpiring, we hold to be the chief cause of the discrepancy.

On the other hand, be the neutral principle ever so abundant, a faulty method of preparation will eventuate in disappointment to the practitioner. The influences of boiling and evaporation upon the neutral principle have already been fully set forth. Hence it will be seen that the chief active principle, instead of being volatile, and so dissipated by boiling as is generally supposed in the preparation of decoctions, syrups, etc., is, by the process of boiling and evaporating, converted into apotheme, and so altered in its constitution and therapeutic properties. Again, we have the highest authority for stating that many varieties of the Sarsaparilla are of no appreciable medicinal value under any circumstances, and taking into consideration the liability of their being thrown into market, we have another fact accounting for the sometimes negative value of the drug.

Thus it will be seen that the divided sentiments of the profession relative to the medicinal value of the Sarsaparilla had each good foundation. That it has proved of positive curative value in many cases and types of disease, is not to be disputed; while it is equally true that it has proved inefficient in a large number of instances. We are induced to believe that the explanations we have given in this volume will reconcile, in a measure, the contrariety of sentiment existing respecting the medicinal value of the plant under consideration, as well as of many others. We believe that the preparations of Sarsaparilla here treated of embody all of medicinal worth pertaining to the plant, and in a form at once concentrated, positive, and uniform in therapeutic character. So far as we have employed the Smilacin, we have every reason to be satisfied with its operation. The precise manner in which its remedial influences are brought to bear upon the system, is a question difficult of solution. That it is alterative and resolvent in its action is manifested by the improvement following its exhibition in those cases in which we know an alterative and resolvent influence to be indispensable. It is not an evacuant, no exaltation of the functions of one organ over another being discernable. It seems to impart a healthful stimulus to the entire glandular system, promoting equally the functions of absorption, secretion, assimilation and depuration. It will at once be seen, therefore, that it is highly restorative in its properties, and peculiarly appropriate in the treatment of various cachexies.

It is scarcely necessary for us to specify the individual types of disease in which the Smilacin may be employed with advantage, yet we will give the results of our observations in a few cases. In scrofula, attended with feeble digestion and an anemic habit, it will be found highly useful. It may be given in doses of from TWO to FIVE grains three times a day. As a general thing, it should be alternated with suitable tonics, as Fraserin, Iron, etc. When deemed advisable, it may be combined with other alteratives. In cold and indolent conditions of the system its operation may be rendered more prompt by combining it with Xanthoxylin, Sanguinarin, Macrotin, or other stimulants.

In the treatment of scrofula, the Smilacin may be depended upon as an efficient auxiliary, if not as an exclusive remedy. We have lately prescribed it in a case of spinal curvature occurring in a patient having a strumous diathesis, and with the most beneficial results. The general health of the patient has been steadily improving since the medicine was commenced. We have noted several cases in which a gradual and steady improvement of the constitutional health has followed the exhibition of the Smilacin. We value it highly in the treatment of rachitis and other diseases of children connected with feeble nutrition. For the purposes of an alterative, resolvent, and detergent, it may be advantageously employed in the treatment of skin diseases, necrosis, caries, and other affections of the bones, ulcers, and for the correction of all morbid cachexies. In syphilis, joined with Irisin or Phytolacin, or Stillingin, etc., the practitioner will find ample opportunity for its employment. Predicating our opinion on the well known remedial value of the plant, when its therapeutic constitution has not been impaired by age, method of preparation, etc., together with a limited experience in the employment of the

Smilacin, which we believe to be its concentrated equivalent, we do not hesitate to recommend it for all the purposes for which the plant has been found useful. Dose, TWO to FIVE grains.

## CON. TINC. SMILAX SARSAPARILLA.

IDENTICAL with the Smilacin, and employed for the same purposes. Average dose, FIFTEEN drops.

# CERASEIN.

Derived from *Cerasus Virginiana.*
Nat. Ord.—*Drupaceæ.*
Sex. Syst.—*Icosandria Di-pentagynia.*
Common Name.—*Choke Cherry.*
Part Used.—The *Bark.*
No. of Principles.—*five,* viz., *resinoid, neutral, amygdalin phloridzin,* and *picrin.*

Properties.—*Tonic, anti-periodic, diaphoretic, febrifuge, anti-spasmodic, and slightly astringent.*

Employment.—*Intermittent and other fevers, debility, indigestion, chorea, hysteria, spermatorrhea, passive hermorrhages, chronic cough, the convalescing stages of diarrhea, dysentery, etc.*

THE Cerasein is one of the most important and valuable acquisitions made to the materia medica of late years. It supplies a necessity long felt by practitioners for a substitute for Quinine in certain conditions of the system wherein the latter is inadmissable. We do not offer it as a complete substitute for Quinine, but as its equivalent in a majority of cases, and as a competent substitute when the latter is contra-indicated. In our own practice we have not prescribed a particle of Quinine in the past two years, having relied upon the Cerasein, in connection with appropriate auxiliary remedies, in the treatment of intermittent forms of disease, and with invariable success. Yet we do not recommend it as adapted to the peculiarities of periodic diseases in every section of the country, well knowing that local influences so modify the action of medicines as to frequently render them of negative value. The existence of these local influences, together with the peculiarities of organization, will forever exclude the discovery of constitutional specific remedies. Nevertheless, we may ascertain a remedy to be possessed of specific therapeutic properties, reliable when the conditions regulating its successful administration are present.

Cerasein is an anti-periodic tonic of remarkable and extended utility. It neither produces cerebral excitement nor deranges the stomach or bowels: but, on the contrary, is a nervine and anti-spasmodic, allaying irritability and quieting the action of the nervous system, and correcting the diarrheal disturbances so characteristic of intermittent fevers. In addition, it is diaphoretic and powerfully febrifuge. Under its influence the skin becomes moist, soft, and flexible, and the pulse, when excited, is reduced in force and frequency, and becomes soft and regular. Upon the mucous membranes of the stomach and bowels it acts in a most desirable manner, deterging morbid exudations, allaying irritability, and restoring the secreting power. It seems to operate remarkably well as an alterative, resolvent and tonic upon the capillary system, hence its utility in passive hemorrhages, night-sweats and other colliquitive and exhausting discharges.

We have employed the Cerasein with uniform success in the cure of ague and fever. The first case in practice in which we had occasion to try it was of the double quotidian type, and of eighteen months duration. We premised **our treatment** by the exhibition of the following powder:

℞.
 Podophyllin........................... gr. j.
 Gelsemin............................gr. ss.
 Asclepin..............................gr. ij.

Mix. This powder was administered in the evening, during the febrile paroxysm, and the use of the Cerasein was commenced next morning in doses of about TEN grains, repeated once in three hours, and so continued for forty-eight hours, then at intervals of four hours for forty-eight hours longer. The dose was then diminished to about FIVE grains, at which quantity it was continued for a few days longer, and such was the success of the treatment that not a single paroxysm of the disease was experienced from the time of taking the first dose, and the patient remains well at the present time, some two years having elapsed since she came under our professional care.

The second case in which we employed it was of the quotidian type, and most inveterate in its character. The patient experienced severe pain in the head upon the approach of every chill, together with irritability of the stomach, nausea, griping pain in the bowels, and a troublesome diarrhea. The Cerasein accomplished a cure in three days. In many chronic cases we have employed the Cerasein with entire success. We remember one case, a lady, who had been afflicted with chills and fever eight months out of the twelve, for four years. The Cerasein, in connection with Podophyllin, effected a permanent cure. But we need not multiply instances to prove its efficacy. The experience of many besides ourself will confirm all that we claim for it.

Much will depend upon the judicious employment of the Cerasein, as regards time, quantity, repetition, continuance, and other necessary conditions, in order to reap success. Our conception of an anti-periodic tonic remedy is, that it is a means calculated to *maintain* a condition, and not to *make* it. In all diseases of a periodic type there is a season of what we might term *comparative health.* It is this condition which we desire to prolong to an indefinite period, and thus render permanent. We should ascertain, therefore, whether the existing condition be one which it would be desirable to confirm, before we employ means to render it permanent. We are of opinion that much mischief is done by the *ill-timed* employment of remedies.

If the condition indicating the employment of anti-periodic remedies does not exist, we must use proper measures to induce it. If there be aberation of the functions of the liver, skin, or kidneys, they must be corrected. Obstructions and morbid accumulations must be removed, the plasticity of the blood obviated, secretion, absorption and depuration established upon a physiological basis, and a condition so brought about, which, if then confirmed, will constitute the accomplished object of sanative medication. It is true that the Cerasein possesses other than anti-periodic properties, all of which are desirable in connection with such a power, but which will not be sufficient, in a majority of cases, to induce the condition we desire to render permanent. Hence we must resort to other remedies, selected with a view of meeting the existing necessities. Thus, if there be hepatic derangement, we have Podophyllin, Leptandrin, Euonymin, Juglandin, etc. As resolvents, we have Veratrin, Sanguinarin, Asclepin, etc., which are also febrifuge and diaphoretic. Stimulants we find in Xanthoxylin, Macrotin, Oil of Capsicum, etc. Gelsemin and Lobelia will supply the relaxant, anti-spasmodic, and other appropriate powers. Thus we need be at no loss for agencies to bring about any condition desired.

The average dose of the Cerasein is FIVE grains, but may be increased to TEN, and even FIFTEEN grains with safety and advantage. The frequency of repetition must be regulated according to circumstances. As a general thing, we find three hours an appropriate interval in the treatment of intermittent fever. The medicine should be continued for some days after the disease is arrested, in order to give tone to the system, and so guard against a return. We have employed the Cerasein successfully in the treatment of intermittent fever occurring during pregnancy. We deem it the safest remedy that can be exhibited, having used it in cases in which the patients were within one month of the period of confinement.

But it is not alone in intermittent fevers that the Cerasein has proved of eminent utility. Remittent, typhoid, and other fevers afford indications for its favorable employment. Being devoid of irritant properties, its employment is admissible in many cases in which other tonics are contra-indicated. Possessing the additional properties of a diaphoretic, febrifuge, nervine, anti-spasmodic and diuretic, its range of application is widely extended. In the convalescing stages of acute diseases we have found it a remedy of great value. It allays irritation, promotes digestion and assimilation, while its diaphoretic, anti-spasmodic and diuretic properties are calculated to fulfill other existing indications. It is for these reasons highly useful for giving tone to the stomach and bowels following an attack of diarrhea, dysentery, or cholera infantum. For the latter purpose it may be advantageously joined with Fraserin.

We have found the Cerasein useful in the treatment of dyspepsia, particularly when there is a tendency to acidification of the food. From FIVE to TEN grains, administered in a little water will generally give prompt relief to that distressing symptom known by the name of heart-burn. When joined with Cornin, or Juglandin, it will prove more efficacious still. Its employment is admissible both in atonic and sthenic conditions of the stomach.

We have employed the Cerasein with much advantage in the treatment of spermatorrhea. We use it in connection with Gelsemin. Our plan of treatment is to exhibit the Gelsemin in proper doses and at suitable intervals until a remission of the symptoms is induced, and then to commence the use of the Cerasein in doses of TEN grains three times a day, exhibiting a dose of Gelsemin at bed time. When deemed expedient, the Gelsemin and Cerasein may be combined. We sometimes combine the Cerasein with Lupulin in this complaint, and with good effect. Further remarks upon the treatment of spermatorrhea will be found under the heads of Gelsemin and Lupulin.

Chronic coughs have been relieved and cured by the use of the Cerasein. General debility, night sweats, and defective circulation, also improve under the influence of the Cerasein.

Passive hemorrhages have also been successfully treated with this agent. When necessary, it may be joined with more powerful styptics and astringents, as the Trilliin, Lycopin, or Oil of Erigeron.

Cerasein has been found of remarkable efficacy in the treatment of herpes, and other forms of chronic febrile exanthema. It breaks up the tendency to periodical eruptions, and effectually obviates the sthenic diathesis. Employed in connection with the Oil of Populus externally, it will effectually cure many cutaneous affections.

Cerasein has also been used with advantage in chorea, hysteria, convulsions, and other affections indicating the employment of an anti-spasmodic and anti-periodic tonic. The full range of employment of the Cerasein is not yet understood, **but we predict for it a steadily extending field of utility.**

# COLLINSONIN.

---

Derived from *Collinsonia Canadensis*.
Nat. Ord.—*Laminaceœ*.
Sex. Syst.—*Diandria Monogynia*.
Common Names.—*Hardhack, Stone Root, Ox Balm, Knot Root, Healall, Rich Weed*, etc.
Part Used.—*The Root*.
No. of Principles—*two*, viz., *resin* and *neutral*.
Properties.—*Tonic, astringent, diaphoretic, alterative, resolvent, and diuretic*.
Employment.—*Diarrhea, dysentery, gout, gravel, dropsy, catarrh of the bladder, leucorrhea, hemorrhoids, colic, cramps, indigestion*, etc.

ALTHOUGH the active principles of this plant have been but recently introduced to the profession, they have rapidly gained well merited favor, and the Collinsonin is entitled to a prominent place in our materia medica.

Collinsonin possesses the therapeutic properties above attributed to it in an eminent degree. It also seems to be entitled to the appellation of carminative, anodyne, and anti-spasmodic, as it expels wind, relieves pain, and relaxes spasm. The sanative influences of the Collinsonin are particularly directed to the absorbent system and mucous membranes. It seems to possess efficient alterative and resolvent properties, and proves efficacious in diseases of the glandular system. In diseases of the bowels and rectum, it stands unrivalled. We have experienced its sanative influences in diarrhea in our own person, and can highly recommend it as a most desirable auxiliary agent in the treatment of all bowel disorders. It soothes, deterges, heals, and gives tone to the intestinal mucous surfaces.

The average dose of the Collinsonin is TWO grains. In diarrhea, dysentery, and cholera infantum, this dose may be repeated once in two hours, except in the latter complaint, in which the dose must be proportioned to the age of the patient. The quantity may be increased or diminished, relatively, according to the urgency of the symtoms in the different affections. When stimulants are indicated, it may be joined with Xanthoxylin, which combination we have employed with much advantage. When astringents are required, it will operate well in connection with Geranin. Combined with Dioscorein, no better remedy can possibly be had for the relief of cramp in the stomach, flatulent and bilious colics, cholera morbus, borborygmus, and all spasmodic affections of the stomach, bowels, and urinary apparatus. For gravelly affections it may be joined with Populin, Senecin, etc.

The Collinsonin has been found highly useful in dropsy, by reason of its peculiar stimulating influences upon the absorbent system. In languid and atonic conditions of the system, it is particularly beneficial, arousing an action in the venous, absorbent and lymphatic vessels, and greatly promoting renal depuration. At the same time it quickens the activity of the cutaneous functions, and, aided by warm diluent drinks, powerfully promotes diaphoresis. It may be employed in connection with Sanguinarin, Ampelopsin, Veratrin, Digitalin, etc.

Collinsonin will be found valuable in the treatment of indigestion, particularly when of an asthenic character, with a tendency to gastritis.

Leucorrhea, catarrh of the bladder, and other critical and excessive mucous discharges may be successfully treated with the Collinsonin, in connection with suitable auxiliary remedies. In these complaints it will be found to answer an admirable purpose in connection with Hydrastin.

But the most remarkable influences of the Collinsonin are observable in hemorrhoids and other diseases of the rectum. The most inveterate and chronic cases are relieved and frequently cured by means of this remedy alone. It should be given in large doses at first, say FIVE grains, and repeated every two hours, in severe cases, until the system is brought under its influence and the symptoms controlled, and then continued in average doses three or four times a day until the disease is eradicated. We have known it to act promptly in suppressing hemorrhage from the bowels, and in relieving those distressing pains characteristic of hemorhoidal affections. It is a valuable constitutional remedy in many affections, and its persevering use seldom fails to benefit the general health. It increases the appetite, and promotes digestion and assimilation.

## CON. TINC. COLLINSONIA CANADENSIS.

Equivalent in properties and employment to the above. Average dose, FIFTEEN drops, increased to thirty in severe cases. We have employed it in connection with the saturated tincture of Xanthoxylum berries, in the treatment of diarrhea, and with excellent effect. Also for pain in the stomach and bowels, etc.

# WINE TINC. LOBELIA INFLATA.

Derived from *Lobelia Inflata.*
Nat. Ord.—*Lobeliaceœ.*
Sex. Syst.—*Pentandria Monogynia.*
Common Names.—*Indian Tobacco, Emetic Weed,* etc
Part Used.—The *Herb.*
No. of Principles.—*Two,* viz., *alkaloid,* and *neutral.*
Properties.—*Emetic, diaphoretic, expectorant, nervine, anti-spasmodic, diuretic, resolvent, and relaxant.*
Employment.—*Croup, pneumonia, bronchitis, hooping cough, asthma, influenza, catarrh, hysteria, chorea, convulsions, poisoning, suspended animation, tetanus, false labor pains, sick-headache, epilepsy, neuralgia, febrile diseases, cutaneous eruptions,* etc.

This preparation of the Lobelia has long been a favorite remedy with us in private practice, and its introduction to the profession has given general satisfaction. The plant yields a number of proximate active principles, but its chief excellences reside in the alkaloid and neutral constituents. These principles are soluble in water, possess the emetic, diaphoretic, expectorant, nervine, anti-spasmodic, diuretic, and relaxant properties of the plant in an eminent degree, and operate without the slightest irritation.  Besides the alkaloid and neutral principles, the Lobelia yields a soft resinoid or oleo-resinous principle, more valuable as an external application than for internal administration.  This oleo-resin is possessed of powerful relaxant properties, and is sometimes administered internally in cases of spasm, convulsions, asthma, and whenever such a property is indicated.  It is this active constituent of the Lobelia that produces the "alarming symptoms" of early writers, and which has caused the Lobelia to be regarded by many as narcotic and dangerous.  But its chief utility is confined to its external employment.  Dissolved in alcohol, it is applied to contracted joints, to the throat in spasm of the glottis, and whenever a powerful relaxant application is needed. In the preparation of the Wine Tinc. this principle is separated from the alkaloid and neutral, and the latter are then redissolved in malaga wine.  The seeds yield a fiixed oil, which will be treated of under the proper head.

The Wine Tinc. is employed for all the purposes of an emetic.  The dose will vary from TWO DRACHMS to TWO OUNCES, and even more in particular cases.  We have had a clinical experience of fifteen years in the use of Lobelia in substance, infusion, alcoholic and acetic tincture, etc., but we give preference to the Wine Tinc. over all other preparations. It is the safest and most reliable emetic, under all circumstances, that can possibly be exhibited.  We are governed in its exhibition, not by the quantity administered, but by the effects produced.  The secret of success is, to give *enough*. It is not uncommon for us to administer from FOUR to SIX ounces of the Wine Tinc. at one time, in the treatment of convulsions, tetanus, etc.  When the tincture cannot be given by way of the mouth, in consequence of the patient's inability to swallow, the quantity intended to be exhibited should be doubled and administered by enema.  Emesis can as readily be produced with the Lobelia employed in this manner as if it were taken in the stomach. It should be diluted with a proper quantity of warm water, and, in some instances, a stimulant joined with it, as the Myricin, Oil of Xanthoxylum, Capsicum, etc. In cases of suspended animation by drowning, hanging, etc., this is the only way in which the medicine can be brought to bear. The following formula may be observed in the above cases, as well as in cases of poisoning, asphyxia, etc.:

℞.
 Wine Tinc. Lobelia............ ℥ VI. vel. X.
 Oil of Capsicum.............gtt. X. vel. XX.

Mix and administer at once with a suitable sized syringe. It would, perhaps, be better to dissolve the Oil of Capsicum in a little alcohol before adding it to the tincture of Lobelia. From ONE to TWO drachms of the tincture of the Oil of Capsicum may be employed, as given under that head. Or, when neither are at hand, ONE drachm of powdered Capsicum may be used instead. This injection should be repeated at suitable intervals until relief is afforded, or until no chance for resuscitation remains. We have known the most desperate cases of suspended animation by drowning to be restored by this treatment when all other means had failed.

In cases of poisoning, particularly when ignorant of the character of the substance swallowed, emetics should never be administered by way of the mouth, but by injection. By neglect of this precaution it frequently happens that the emetic is neutralised and does not operate, either in consequence of chemical reactions, or from paralysis of the nerves of the stomach. There is, also, a liability to the formation of dangerous compounds by the mutual reactions which take place between the substances introduced and the substances already there. These remarks apply when the character of the poison swallowed is not known. One very essential condition to be observed in connection with the employment of the Wine Tinc. of Lobelia as an emetic, either per os or per anum, is, that undue acidity of the stomach and bowels be neutralised, either by the previous administration of an alkalie, or by combining it with the Lobelia when exhibited. It frequently happens, when this precaution is neglected, that the emetic influences of the Lobelia are suspended, and the medicine passes off by perspiration, stool, and urine. Acids effect a destructive decomposition of the neutral principle, and hold the alkaloid in solution, thus suppressing its action. Tannic acid is incompatible with the alkaloid principle, forming with it an insoluble compound, and thus rendering it inert. When soda or other carbonic alkalies are administered for the purpose of neutralising acidity, severe pain will be experienced in the region of the stomach, accompanied with a death-like nausea. This is occasioned, probably, by the sudden disengagement of the carbonic acid of the alkalie, the base combining with the lactic or other acids present. It does not occur, however, in every instance, and is relieved as soon as vomiting takes place. When this phenomenon is properly understood, it prevents unnecessary alarm on the part of the patient. If the precautions here noted in regard to neutralising acidity be neglected, the Lobelia will be very tardy in manifesting its emetic influences, and, in many instances, will not operate at all.

As a remedy in the treatment of mucous and spasmodic croup, the Wine Tinc. of Lobelia is superior to any other single agent. Its purely innoxious character renders it a safe and reliable remedy for patients of all ages, from the infant to the septagenarian. In the management of this disease the Lobelia must be administered promptly and in full doses, and repeated at intervals of from ten to thirty minutes until free vomiting ensues. It is necessary to induce complete relaxation of the system by means of full emetic doses, and afterwards to maintain it with smaller doses repeated at suitable intervals. When inconvenient or difficult to administer it by the mouth, as in the case of infants and children, it should be given by injection. The same directions will apply in cases of pneumonia, asthma, convulsions, hysteria, tetanus, etc. In croup the Lobelia is sometimes joined with Sanguinarin, and with advantage. In other cases with Eupatorin Perfo. We have seen the Lobelia employed to a considerable extent in the treatment of pneumonia, and with the happiest results. We remember the case of our little sister, who, at two years of age, was attacked with this complaint, and to whom six Lobelia emetics were administered daily for several consecutive days, and we believe them to have been the means of saving her life. Lobelia not only unloads the lungs of the accumulated secretions, but it also resolves the plasticity of the blood, relaxes spasm, promotes diaphoresis, and changes the entire diathesis of the system. In all febrile disorders manifesting a determination to the brain, or a tendency to congestion, we have, in the Lobelia, one of the most reliable derivative remedies yet discovered. Here its powers of relaxing constriction, equalising the circulation, promoting absorption, secretion and exhalation are particularly called for, and will seldom disappoint the practitioner. The necessities of particular cases will best indicate the manner of employing the Lobelia. If it be desirable to produce sudden revulsion, as in severe and sudden congestions, it should be exhibited in full emetic doses, say from ONE to THREE ounces. In other instances, broken doses frequently repeated will subserve a better purpose. The latter plan of administration should be adopted in the treatment of low delirium, tonic spasm, and febrile disorders generally. In confirmed and lingering cases of typhoid fevers this course will be found of much service. A case occurred in our practice in the fall of 1846, when typhoid fever was prevalent, in which we administered one drachm of the infusion of the Lobelia herb every hour in the twenty-four for eight days consecutively, and we believe it to have been the means of effecting a cure. The patient had a rapid convalescence, and "still lives." The fever had been running eleven days before we were called.

One noticeable feature in connection with the operation of Lobelia, as an emetic, is this, it does not derange the functions of digestion. In the treatment of chronic diseases, the patient, after having been subjected to the operation of a Lobelia emetic, is enabled, in thirty or sixty minutes thereafter, to eat his dinner, and not only eat but digest it. In the treatment of indigestion, the exhibition of a Lobelia emetic has frequently enabled the patient to eat and digest a substantial meal, whereas he had not been able to either receive or retain food upon the stomach for a considerable time. Its sanative influences, in many instances, seem to be almost electrical. We would mention, in connection with this idea, that, while under the influence of a Lobelia emetic, the patient frequently experiences a sensation as if a strong galvanic current was passing through the system, or rather the stomach seems to be the centre from which radiate numerous currents, passing along the limbs and to the periphery of the entire nervous system. These sensations resemble a series of rapid galvanic shocks, accompanied with a feeling of numbness, and pass off with the operation of the medicine.

In the treatment of chorea we give an emetic of the Wine Tinc. of Lobelia every other day, or every day in severe cases, and alternate with Cerasein, Hydrastin, Cornin, Capsicum, Scutellarin, Gelsemin, and other tonics and anti-spasmodics.

In spasmodic asthma we administer the tincture in quantities sufficient to relieve the immediate symptoms, and then continue the same in suitable doses, and at proper intervals, in connection with appropriate auxiliary remedies, until a cure is effected. We observe the same method in the treatment of influenza, hooping cough, and other affections of the respiratory organs. Ordinary catarrh or cold in the head may be relieved by taking from FIVE to TEN drops of the undiluted tincture at a time, and repeating as occasion requires. The benefit derived is more in consequence of the stimulating effect of the Lobelia upon the glands of the throat, than from its passing into the stomach. For the colds, coughs, and "snuffles" of children, we mix the tincture with molasses or sugar-house syrup.

℞
Wine Tinc. Lobelia.................... ʒi.
Sugar-house Syrup.................... ʒiss.

Mix. Dose, from one half to one teaspoonful every hour or

two. This will be found excellent for ordinary coughs and colds. Of course, the dose must be varied to suit the occasion. As an expectorant, the Lobelia has few equals, and no superior. It is of much utility in pleuritis, overcoming the viscidity of the pulmonary secretions and favoring expectoration.

The Wine Tinc. of Lobelia is a remedy of great value in the treatment of disorders of the female system. We have already spoken of its remarkable efficacy, in connection with Myricin, in relieving spasmodic and false labor pains. The reader is respectfully referred to the article on Myricin for a description of the method of employing it, and thus save us the necessity of repetition. Equally efficient will the Lobelia be found for controlling undilated and undilatable os uteri, puerperal convulsions, puerperal fever, retention of the placenta, etc. Our method of employing it in the latter instance is by injection per anum.

℞.
    Wine Tinc. Lobelia........................ ℨ ss,
    Cypripedin ..........................grs. X.
    Warm Water........................ ℨ IV.

Mix. Administer blood warm, and repeat once in thirty minutes, if found necessary. The efficacy of this remedy in promoting the expulsion of retained placenta needs to be witnessed in order to be fully appreciated. The same injection will be found of great service for the relief of pains attendant upon the passage of calculi through the ureters, and for suppression and retention of the urine. In the latter affections its efficacy will be materially enhanced by the addition of from TEN to FIFTEEN grains of Myricin to each enema.

An occasional emetic of the Wine Tinc. of Lobelia is frequently of great service in the treatment of diarrhea and other intestinal disorders. In cholera morbus, when the stomach is loaded with acrid ingesta, it should not be omitted. It has been employed in asiatic cholera, in the same conditions, with most excellent effect. Prolonged nausea and vomiting depending upon spasm of the stomach are effectually relieved with broken doses of the tincture. In ordinary cases the tincture may be reduced with water when used as an emetic, but in urgent cases, and where smallness of dose is an object, we administer it without admixture. In croup and convulsions of children it is better administered undiluted.

For relieving the ill effects of drinking too freely of cold water while heated, there is no better remedy than the Wine Tinc. of Lobelia. It should be given in large and repeated doses, and continued until complete reaction is established and the circulation equalised. In some cases it may be advisable to combine a stimulant with the Lobelia, for which purpose we prefer the Capsicum or its preparations to anything else.

A Lobelia emetic will give speedy relief in cases of sick-headache, and, where they are chronic, its occasional repetition will frequently break up the constitutional diathesis. Neuralgia is often relieved by the same means.

For relaxing constriction and favoring the development of the eruption in exanthematous fevers, we have, in the Lobelia, a most excellent remedy. The doses and repetitions must be governed by the necessities of the case.

Externally, the tincture is applied in cases of erysipelas, various eruptions, and, diluted with water, is employed in the treatment of purulent, strumous, and other forms of opthalmia. Also to the throat and chest in croup, asthma, etc.

Finally, the Wine Tinc. of Lobelia may be employed with advantage in all spasmodic affections, and whenever an emetic, nauseant, diaphoretic, anti-spasmodic, expectorant, or relaxant is indicated. It is neither narcotic nor dangerous, and may be employed with perfect safety for fulfilling any of the indications embraced within its range of therapeutic properties. Experience in its employment will confirm the confidence of every practitioner in its utility, and he will learn to look upon it as an indispensable agent of the materia medica. We are far from deeming it a specific, yet we hold it capable of fulfilling specific indications with far more certainty and safety than any other remedy, and one for which there is no substi tute. We should feel lost without it, and are confident that such would be the expression of all who become acquainted with its true value.

# OIL OF LOBELIA.

---

DERIVED from the seeds of the *Lobelia Inflata*.

The oil of Lobelia is chiefly valued as an expectorant, antispasmodic and relaxant. Although sometimes used for the purposes of an emetic, it does not operate so kindly as the preparation first treated of. Internally, it is employed with much benefit in the treatment of asthma and other affections of the respiratory organs. The medium dose of the oil is ONE drop, repeated three or four times daily. It may be administered on sugar, or suspended in mucilage. It will be found a valuable expectorant and relaxant, and may be employed with advantage in all spasmodic affections. It may be combined with other agents at the pleasure of the practitioner.

But it is in combination with the Oil of Capsicum that we make most use of this agent. Our formula is as follows:

℞.
  Oil of Lobelia........................
  Oil of Capsicum.....................aa. ℨi,
  Alcohol, 95 per cent.................. ℨii.

Dissolve the oils in the alcohol and it is ready for use. The dose of this preparation is from FIFTEEN to SIXTY dops. We employ it in apoplexy, asphyxia, convulsions, suspended animation, asiatic cholera, tetanus, and all violent spasmodic disorders. No physician should be without this remedy at hand. In cases of fainting, falls, concussions, drinking too freely of iced-water, violent spasmodic pains in the stomach and bowels, and whenever it is necessary to relax spasm, equalise the circulation, and so bring about a re-action, this remedy is unequalled. When it cannot be swallowed, the quantity may be doubled and administered by enema. In tetanus, when the jaws are set together, also in hysteria, and other convulsions, a quantity of this preparation poured between the teeth, will, as soon as it reaches and has time to act upon the muscles of the throat, relax the spasm and enable the patient to open his mouth and swallow. At the same time the throat may be bathed externally with the same. Neuralgic and rheumatic pains, toothache, etc., are relieved by bathing with this preparation. When the tooth is decayed and the nerve exposed, it may be applied on cotton. We have treated many cases of apoplexy with this medicine, and, in connection with hot mustard foot baths and the application of cold to the head, with invariable success.

The oil applied to the throat externally has given prompt relief in spasm of the glottis, croup, etc., and applied to the chest relieves dyspnea. It enters into the Comp. Stillingia Liniment, for the formula of which see Oil of Stillingia. In applying the oil to infants and children, externally, care must be taken not to apply it too freely, as more relaxation may be produced than is desirable, together with nausea and vomiting

In spasmodic croup, the oil may be given in doses of ONE drop, and repeated once in thirty minutes until relief is afforded. But we prefer the Wine Tinc. in the treatment of the disorders of infants and children.

# CON. COMP. STILLINGIA ALTERATIVE.

**FORMULA.**

Rad. *Stillingia Sylvatica.*
" *Corydalis Formosa.*
" *Phytolacca Decandria.*
" *Iris Versicolor.*
Cort. *Xanthoxylum Fraxineum.*
Fol. *Chimaphila Umbellata.*
Sem. *Cardamomum.*

WE quote from the manual of Messrs B. Keith & Co., the following extracts explanatory of the character and peculiarities of this preparation:

"Complaints having reached us that the above syrup, (Syr. Stillingia Comp.) as put up by manufacturing druggists, had failed in numerous instances of exercising its accustomed remedial influences, we directed our attention to the discovery of the cause, and the remedy. The former we found to depend upon the fact that, in its preparation, the starch, grape-sugar, and the other non-medicinal elements were retained, and in consequence of there not being alcohol enough present to resist a tendency to fermentation, a destructive chemical decomposition ensued, whereby the therapeutic elements were destroyed. Syrups long made up, undergo a progressive disintegration of their therapeutic constituents, and thus become unreliable and unfit for use."

"Another pertinent reason is found in the fact which we have heretofore advanced, that is, the uncertain amount of active principles any given number of pounds of a crude article will yield. Hence, so long as organic pharmaceutic compounds are regulated by the *weight of the crude substances* of which they are composed, instead of the *actual amount of active principles* present, there can be nothing but uncertainty in regard to their medicinal strength."

"Having ascertained, by repeated analyses, the utmost yield of the above articles, when dictated by weight, we are no longer governed by a stipulated number of pounds, but by the *actual product of active principles*. Our estimate is based upon *therapeutic* and not upon *physical* considerations. In this way we secure an uniformity in no other way attainable, and avoid the discrepancy in remedial value which renders ordinary syrups unreliable."

"One ounce of our preparation is equivaelent to 32 ounces of the Comp. Syrup of Stillingia as prepared by other druggists, when of maximum strength."

'The dose of the latter is from one fluid-drachm to one fluid-ounce.'"

'The dose of our preparation is from two to five drops.'

"Any practitioner so inclined, may prepare one quart of Comp. Stillingia Syrup in a few minutes, by adding an ounce of our Con. Comp. Stillingia Alterative to thirty-one ounces of simple syrup, and flavoring as preferred. We warrant our preparation against change in any climate, and for an unlimited period of time."

"Thus are portability, uniformity of strength, convenience of administration, and protection against inertness secured."

The reader may learn, by referring to their respective heads, the properties of the various ingredients composing this preparation, and thus form some conception of its range of applica tion. Although opposed to such complexity of combination, we must acknowledge that our experience in the employment of the Con. Comp. Stillingia Alterative has been of the most gratifying character. With it we have treated scrofula, syphilis, cutaneous eruptions, hepatic disorders, rheumatism, mercurial affections, leucorrhea, gonorrhea, glandular enlargements, and almost every form of disease requiring the employment of an alterative, resolvent, and tonic remedy. As a constitutional remedy in the treatment of contagious, purulent, and strumous opthalmia, when not complicated, it is remarkably efficient, and seldom will any other remedy be needed.

Although the dose of this preparation averages from TWO to FIVE drops, we frequently increase it gradually to TEN, finding cases and conditions requiring more than the average dose to produce the desired effect. We find that it operates much better by exhibiting it two hours after eating, than when given shortly before meals. In the latter instance it interferes with the appetite, and when food is taken nausea is produced, as is also the case when the dose is too large. The best way of administering it is to drop it into a little cold water. It is easily made into syrup, as above stated.

# STRYCHNIN.

Derived from *Strychnos Nux Vomica.*
Nat. Ord.—*Apocynaceæ.*
Sex. Syst.—*Pentandria Monogynia.*
Common Name.—*Nux Vomica.*
Part Used.—The *Seeds.*
No. of Principles.—*three,* viz., two *alkaloids,* (*strychinia and brucia,*) and a *neutral* principle.

WE have never employed the Strychnin in practice ourselves, but the concurrent testimony of those who have tested it clearly defines it to be equivalent to the Nux Vomica in therapeutic properties, and infinitely preferable to the extracts and other preparations of that remedy, as it is of definite, reliable, uniform, and unchanging medicinal strength. It is employed for all the purposes for which the seeds have been found beneficial, for a history of which the reader is respectfully referred to the U. S. Dispensatory, and other standard works on materia medica. It has been manufactured by request, and has given satisfaction to those for whose use it was prepared. It is ONE THIRD LESS in remedial strength than the Strychnine of commerce. Thus, if the dose of the

Strychine be from ONE SIXTEENTH to ONE TWELFTH of one grain, the dose of the Strychnin will be from ONE TWELFTH to ONE EIGHTH of one grain, or thirty-three and one-third per cent. more.

It is a medicine of great power, and will not bear to be incautiously trifled with.

# CON. TINC. STRYCHNOS NUX VOMICA.

THIS preparation is equivalent to the Strychnin above described. The dose is from ONE FOURTH of ONE DROP to ONE DROP. It is simply a solution ot the active principles composing the Strychnin. For a history of its properties and uses, the reader may consult standard authorities upon materia medica.

# CON. TINC. CANNABIS INDICA.

Derived from *Cannabis Indica*.
Nat. Ord.—*Canabinnaceæ*.
Sex. Syst.—*Diœcia Pentandria*.
Common Name.—*Indian Hemp*.
Part Used.—The *Herb*.
No. of Principles—*two*, viz., **resinoid**, and **neutral**.
Properties.—*Narcotic, anodyne, anti-spasmodic, etc.*
Employment.—*Nervous diseases generally*.

WE are not enabled to record our personal experience of the utility of the Cannabis Indica, never having employed it in practice. According to the U. S. Dispensatory "it is recommended in neuralgia, gout, rheumatism, tetanus, hydrophobia, epidemic cholera, convulsions, chorea, hysteria, mental depression, insanity and uterine hemorrhage." Of the *modus operandi* of this remedy we have been enabled to learn but little. So far as we can ascertain, it is of doubtful and uncertain effect, its administration being attended with great disparity of action. The Dispensatory further tells us, "in morbid conditions of the system, it has been found to produce sleep, to allay spasm, to compose nervous inquietude, and to relieve pain." These, we apprehend, are the accidental deductions made from the joint experiments and opinions of various practitioners. Further than this, nothing entitled to our credence has been adduced. For our own part, we do not look upon the Cannabis Indica as a desirable acquisition to our materia medica, much less an indispensible one, as we know of no indications it is capable of fulfilling that cannot be met with other medicines, and with far more precision, certainty, and uniformity of action.

We are of opinion that medicinal plants grown in remote sections of the earth, and known to produce certain specific physiological influences upon the natives of that locality, should not be looked upon as being capable of inducing the same train of results when transferred in their application to the people of another clime. Differences of organization, temperament, habits, occupation, diet, climate, and other influences all tend to modify the impressibility of the nervous system, and correspondingly will the means of therapeutic impression vary in their operation upon the living forces. We hold that the experiments of Dr. O'Shaughnessy in India cannot be accepted as a criterion in estimating the remedial value of the Hemp in this country. In a volume entitled "Headland on the action of medicine," the reader may find recorded some interesting information in regard to the diversity of therapeutic action. In speaking of opium, he tells us, "in the Caucasian race it generally produces somnolency; in the Chinese, intoxication; in the Javenese, and Malays, it will cause a raving delirium." And from some notes of an intelligent reader and writer we take the following:—"Do not ardent spirits act in the same mysterious way upon the different races? It is seldom that an Indian becomes "jolly"—he is, as a rule, sullen, morose, and savage. The Negro is sleepy. The Malay is a raving, blood-thirsty maniac." These facts would seem to support the conclusions we have come to in the preceding paragraphs. When uniformity of organization and temperament shall become a national characteristic, then may we expect to find the people of that nation similarly exercised by the exhibition of a given therapeutic agent. True, nations as well as individuals have their distinguishing characteristics, but each are subjected to a variety and diversity of modifying influences. In the one instance the phenomena produced are national; in the other, individual. As we see individuals among people of the same race variously impressed by alcohol and other narcotics, so may we behold it of nations. In an individual case opium soothes and depresses; in the other it excites and exhilerates; in a majority of instances it constipates the bowels, while we have known individuals to employ it for the purposes of a cathartic, being freely purged by even a small quantity.

We have adduced the fact, in the first chapter of this work, that the Cannabis Indica grown upon the hills of India is entirely different from that grown in the valleys, an additional evidence of the uncertainty of the plant as a reliable remedial agent. But it is not improbable that time and further experiment may enable us to overcome these objections, and to give, in future editions of this work, a fuller and more reliable history of the remedial value of this plant. The average dose of the tincture is **FIVE drops**.

---

We now conclude our history of the therapeutic properties of concentrated medicines proper, hoping, in future editions, to enlarge the list by making such additions as the necessities of the profession demand. The Erythroxylin, from the Erythroxylum Coca, and the Daturin from the Datura Stramonium, are now under consideration, and as soon as they shall have been thoroughly tested in clinical practice, the history of their therapeutic properties and range of employment will be laid before the profession. The Con. Tinc. Gossypium Herbaceum is likewise being put to practical tests, but so far the results secured have not been sufficiently definite to enable us to recommend it to the confidence of the profession.

We are conscious that we have not embraced the entire range of application of the various remedies described in this work, yet we have endeavored faithfully to portray their therapeutic action. Since penning the article on Gelsemin, we have employed that agent extensively in the treatment of bowel disorders, and with the most satisfactory results. For controlling the spasmodic action of the stomach and intestinal tube, it far excels any single remedy we have yet employed. It soothes the irritability of the mucous surfaces, and completely controls the spasmodic tendency. For the relief of tenesmus, we employ the Con. Tinct., adding from TEN to SIXTY drops to an enema, according to the severity of the case and the age of the patient, and repeat as occasion requires. It operates admirably. The Con. Tinc. Senecio Gracilis has been found, by several practitioners, an excellent and reliable remedy for allaying the nausea attending pregnancy. The Con. Tinc. Gelseminum has been applied with complete success to counteract the effects of the bite of a spider, relieving the pain, abating the inflammation and swelling, and effecting a cure. So we might go on enumerating instances of the diversified application of these remedies, but space will not admit of a lengthy recapitulation, and we shall be content to submit the question of adaptation to the intelligent judgment of our readers, trusting that our feeble efforts to elucidate the history of these agents may shed some light upon their pathway.

# ADDENDA.

## ACONITIN.

Derived from *Aconitum Napellus.*
Nat. Ord.—*Ranunculaceæ.*
Sex. Syst.—*Polyandria Trigynia.*
Common Names.—*Wolfsbane, Monkshood.*
Parts Used.—*Leaves* and *Root.*
No. of Principles.—*Three,* viz., *resin, neutral,* and *alkaloid.*

Properties.—*Diaphoretic, diuretic, alterative, antispasmodic,* and *narcotic.*

Employment.—*Phthisis, dropsy, gout, neuralgia, rheumatism, paralysis, portal congestions, hysteria, etc.*

IN small and frequently repeated doses, Aconitin promotes diaphoresis and diuresis, and increases the secretions of the mucous, serous, and synovial membranes. Its long continued use is attended with the appearance of exanthematic eruptions upon the skin, accompanied with a troublesome itching, and severe pain in the joints.

In larger doses Aconitin gives rise to severe cardialgia, paralysis of the tongue and pharynx, a sense of suffocation, vomiting, painful diarrhea, quick and irregular pulse, dyspnea, swelling of the abdomen, tremors of the limbs, followed in due time by extreme prostration, chills, severe pains in the head, bones and joints. After a longer or shorter duration of these symptoms, the patient is attacked with profuse sweats, together with an increased flow of urine, and oftentimes a measley looking eruption makes its appearance on the skin. Permanent derangement of the digestive functions, together with a jaundiced condition of the system, are the general sequents of excessive doses of Aconitin.

Large doses of Aconitin sometimes prove speedily fatal, preceded by convulsions, delirium, cerebral congestions, tetanus, &c. A post-mortem examination in these cases reveals severe congestion in the veins of the head, lungs, and abdomen. Sometimes, but not always, inflammation of the membranes of the stomach and intestines is present.

The above described dynamical effects of Aconitin demonstrates it to be a stimulant to the nerves of sensation and to the secreting apparatus generally, but more particularly to the veins, skin, kidneys, mucous and synovial membranes, and the sheaths of the muscles and tendons, increasing their secernent activity, and exalting their sensibility and irritability. It also hastens the metamorphosis of the fluidiform materials of the circulation.

From a consideration of the physiological influences of Aconitin, it has been recommended in those forms of disease originating in a suppression of the peripheric secretions, particularly in obstinate chronic cases—also in chronic affections of the sheaths of the muscles, tendons, and nerves —of the fibrous membranes and organs—of the mucous and synovial membranes—for the resolving of exudates and dispersion of swellings in these organs, such as are dependent upon inactivity or obstruction—in paralytic affections of the nerves, and in those neuralgic disorders which originate in local metastastic, rheumatic or arthritic affections of the nenrilema. Aconitin has also been recommended in phthisis pulmonalis, in the incipient stage, beginning with small doses and gradually increasing. Aconitin is contra-indicated in the presence of pneumonic inflammations and congestions, high febrile excitement, and colliquitive sweatings.

Aconitin has been found of benefit in the asthenic forms of dropsy, particularly when arising from suppressed perspiration, rheumatic and arthritic cachexies, and especially when located in the skin and joints. In connection with Podophyllin, Veratrin, Jalapin, Apocynin, &c., Aconitin has been successfully employed in the treatment of portal congestions, and for the correction of those functional derangements of the abdominal viscera manifesting unusual torpor, occurring in individuals of a cold, lymphatic or phlegmatic habit, though contra-indicated when plethora or excessive nervous sensibility of those organs is manifest.

Aconitin has been successfully employed in different forms of rheumatism, even in the acute varieties when the fever and erethism are diminishing, or have entirely ceased. In lingering rheumatic pains of the joints, rheumatic headaches, rheumatic cardialgia, rheumatic metrorrhagia, and obstinate neuralgias, occurring in asthenic habits, Aconitin has likewise proved a valuable remedy. Also in atonic gout, asthma, &c., combined with Asclepin, Eupatorin Purpu., Veratrin, and in cases of great nervous sensibility, with Gelsemin.

To recapitulate the principal uses of Aconitin, we may mention, all that class of diseases arising from or dependent upon suppressed cutaneous or other secretions, or inactivity of the secernent vessels, as rheumatic, arthritic, strumous, syphilitic, psoric, and mercurial cachexies, glandular enlargements, obstinate salt rheum, itch, synovitis, amaurosis, deafness, paralysis, as of the extremeties, bladder, &c., incontinence of urine, &c. The writer has employed the Tincture with much success in the treatment of hysteria, more particularly of the chronic forms, and in the absence of acute inflammations or congestions.

Aconitin is contra-indicated in acute inflammation, hypers

thenic fevers, gastritis, threatened congestions of the brain, lungs, or other organs, colliquitive sweats, great irritability of the nerves of sensation, and acute hepatic affections.

The dose of Aconitin is from ONE TWENTY-FOURTH to ONE TWELFTH of ONE GRAIN.

## CON. TINC. ACONITUM NAPELLUS.

Derivation and properties same as Aconitin. The internal employment is the same. Externally, the Con. Tinc., diluted with eight times the quantity of water, is employed as a collyrium in rheumatic and arthritic inflammation of the eyes. The dose of the Con. Tinc. is from ONE to FIVE drops.

# COLOCYNTHIN.

---

Derived fron *Cucumis Colocynthis.*
Nat. Ord.—*Cucurbitaceæ.*
Sex. Syst.—*Monœcia Monadelphia.*
Common Names.—*Colocynth, Bitter Cucumber.*
Part Used.—The *Fruit.*
No. of Principles.—*One,* viz., *resinoid.*
Properties.—*An irritant hydragogue cathartic.*
Employment.—*Obstinate quartan fevers, atonic jaundice, indolent dropsies, amenorrhea, worms, chronic nervous affections, &c.*

In small doses, Colocynthin accelerates the peristaltic motion f the intestinal canal—increasing the mucous and other secretions; promotes the activity of the abdominal blood-vessels, and quickens the functions of the lymphatic and glandular systems, and of the kidneys. In large doses, Colocynthin gives rise to severe griping pains in the abdomen, vomiting, a violent diarrhea, with frothy discharges, accompanied with tenesmus and hemorrhage of the rectum. In yet larger doses, it gives rise to the same train of symptoms in a more aggravated form, followed by vertigo, blindness, deafness, delirium, convulsions, and death. The fatal effects are produced by excessive and exhaustive irritation, accompanied, in some instances, with gangrene of the rectum. The continued employment of Colocynthin produces, like all other drastic remedies, paralytic-like debility of the bowels and rectum, suppressed secretion, and obstinate constipation.

Colocynthin is employed, in small doses, in excessive torpor of the abdominal organs, particularly of the lymphatics, glands, mucous membranes, and nervous plexus, and in those disorders arising from or supported by said abnormal conditions. Of this class we may mention obstinate and frequently recurring quartan fevers, atonic jaundice, retention of the catamenia and hemorrhoidal discharges, indolent dropsies, ascarides and chronic blenorrhea.

Colocynthin has been employed with some success in the cure of those chronic nervous ailments based upon or supported by a general torpor of the nerves of sensation, or upon local paralysis of the abdominal and lower spinal nerves. It would seem to act, therefore, in the latter instance, as a local deducive stimulant, and, when long continued, as a stimulant to the entire nervous system.

In the treatment of mania, melancholy, epilepsy, chronic nervous vertigo, and headache, Colocynthin is employed in doses sufficient to purge; a considerable interval—say several days—being allowed to elapse between the repetitions of the doses. Small and repeated doses of Colocynthin have proved useful in the treatment of mild forms of mania, lethargies, and as a prophylactic of serous and mucous apoplexies, paralysis of the rectum, urinary organs, and lower extremities. Its employment is contra-indicated, however, in the presence of an inflammatory condition of those organs. Colocynthin has been employed with some success in dyspepsia, arising from a paralytic debility of the stomach and its appendages.

The **dose of Colocynthin will vary from ONE-HALF to TWO GRAINS.**

# RHEIN.

**Derived from** *Rheum Palmatum.*
Nat. Ord.—*Polygonaceæ.*
Sex. Syst.—*Enneandria Trigynia.*
Common Name.—*Rhubarb.*
Part Used.—*The Root.*
No. of Principles.—*Three, a resinoid, and two neutrals*
Properties.—*Cathartic, alterative, laxative, tonic, resolvent chologogue, and anti-septic.*

Employment.—*Dyspepsia and its concomitant symptoms, heart-burn, flatulence, constipation, &c., diarrhea, dysentery, colic, atonic dropsy, chlorosis, mucous cachexies, scrofula, diabetes mellitus, fevers, hemorrhoids, jaundice, biliary calculi, asthenic catarrhs, etc.*

Administered in small doses, Rhein stimulates the digestive apparatus, improves the appetite, promotes the formation of chyle and the supply of bile, and corrects disturbed action of either function.

It exercises a general tonic influence over the secretive functions, and particularly those of the mucous membranes. In very large doses, Rhein gives rise to diarrhea, which is usually followed by constipation.

Rhein, in small doses, may be usefully employed for the relief of heart-burn, flatulence, diarrhea, constipation, and other symptoms attendant upon indigestion, and for the correction of the excessive mucous discharges which sometimes follow an attack of gastric or bilious fever. In asthenic dysenteries, it may be usefully combined with Leptandrin, Collinsonin, Cerasein, Fraserin, &c. It is a valuable remedy in the digestive disorders of children, such as vomitng of the food, colic, diarrhea, and convulsions produced by the retention of acrid ingesta. In the disorders of dentition, it answers an admirable purpose, in combination with alkalies and aromatics.

Rhein is also of much utility in the treatment of chlorosis, leucorrhea, dropsy, scrofula, rickets, diabetes mellitus, and atonic hemorrhoids. For the cure of jaundice, and for the removal of biliary concretions and impacted fœces, the Rhein is said to be of remarkable efficacy. Finally, in all disorders connected with the digestive and assimilative apparatus, either of the organs themselves, or from sympathy therewith, and in all disorders of the mucous surfaces, the Rhein will be found a remedy of much value.

Rhein is sometimes employed locally as an application to foul ulcers, on account of its tonic and antiseptic properties, and in the form of an injection to restrain excessive hemorrhoidal and leucorrheal discharges.

Rhein is contra-indicated in active inflammations, congestions, and hemorrhages.

**The dose of Rhein is from ONE to FOUR GRAINS.**

# ATROPIN.

Derived from *Atropa Belladonna.*
Nat. Ord.—*Solanaceæ.*
Sex. Syst.—*Pentandria Monogynia.*
Common Names.—*Belladonna, Deadly Nightshade, &c.*
Parts Used.—*Leaves and Root.*
No. of Principles.—*Three : resin, neutral,* and *alkaloid.*
Properties.—*Narcotic, anodyne, antispasmodic, calmative, alterative, resolvent, diaphoretic,* and *diuretic.*

Employment.—*Convulsions, epilepsy, neuralgia, schirrus, dropsy, obstinate intermittents, scarlet fever, whooping cough, asthma, suppression of the menses, syphilitic infections, paralysis, amaurosis, nervous affections, mania, melancholy, &c.*

The dynamical effects of Atropin, when given in small doses, are dryness in the fauces, thirst, difficult deglutition, deluded vision, increased sensibility and irritability of the optic nerve, dilated pupil, vertigo, mental exhiliration, and increased perspiration. When the doses are increased, the thirst becomes excessive, swallowing is difficult if not impossible, the throat becomes swelled and painful, with spasm of the glottis, a sense of numbness is felt about the eyes, followed by delirium, mania, hiccough, dyspnea, grinding of the teeth, convulsions, tetanus, lethargic slumber, and apoplectic death. A post-mortem examination reveals severe and extensive congestions of the brain, lungs, liver, spleen, stomach, and intestines. The spleen is soft and easily separated between the fingers, the blood is in a state of decomposition, and the body soon putrefies.

According to the experiments of Orfila, Atropin acts most speedily when taken into the stomach, or injected into the veins; more slowly when brought in contact with the cellular structure.

Important indications for the employment of Atropin are found in the early stages of organic affections, such as induration and schirrus of the more important organs, and in the dispersion of glandular enlargements. The peculiar utility of Atropin in these cases depends, in addition to its alterative, resolvent, and stimulant powers, upon the possession of remarkable anti-spasmodic and sedative properties, whereby it soothes and overcomes the abnormal sensibility giving rise to and accompanying these structural changes. The employment of stimulating resolvents devoid of these auxiliary properties would, under like circumstances, be more likely to aggravate the disorder by provoking the existing irritation to a dangerous extent. The use of Atropin is said to have cured fully developed indurations, even when of long standing; but in general its influence in these cases goes no further than to arrest the development at its present stage, and to act as a prophylactic of cancerous degeneration.

Atropin is employed in the treatment of mania, preceded by the use of alteratives and relaxants. Much care must be exercised in its employment, and all existing idiosyncrasies carefully noted. The encephalic constitution is said to bear this remedy best.

Atropin is likewise said to have been successfully employed in the treatment of hypochondria, hysteria, epilepsy, chorea, and other nervous diseases dependent upon abominal obstructions or suppressions, as of the menses, or upon a morbid exaltation of the nervous sensibility of the parts. It is exhibited in connection with Podophyllin, Veratrin, Lobelia, Rhein, &c. In connection with Lobelia, Podophyllin, Hyosciamin, Prunin, Asclepin, &c., it has been highly recommended in whooping cough, asthma, and other affections of the respiratory system. In various forms of neuralgia, the use of Atropin, both internally and externally, has been attended with much success.

In hydrophobia, Atropin is said not only to act as a preventive, but also to have effected a cure in several instances. In order to be efficacious, it must be given in sufficient doses to induce a degree of narcotism, and its use persevered in. While under its influence the patient will frequently complain of a smarting sensation in the wounds inflicted by the bite.

Paralysis dependent upon torpor of the abdominal functions is said to have been successfully treated with Atropin. In dropsy arising from biliary derangement, this remedy has found useful employment. In suppression or defective flow of the catamenial and lochial secretions, when arising from obstructions in the portal system, and in rheumatic, arthritic and exanthematic metastases, salt rheum, and even in long standing syphilitic infections, Atropin is recommended as a remedy entitled to much confidence. In chronic nervous rheumatism Atropin will afford much relief. It is sometimes employed to prevent abortion in consequence of too great sensibility and contractility of the uterus. Small doses are exhibited at bed time.

Atropin has gained considerable reputation as a prophylactic of scarlet fever, and is also extensively employed in the treatment of that malady.

Contra-indications to the employment of Atropin are, high inflammatory excitement, plethora, tendency to congestion of the brain, lungs, or other organs, erethism of the blood, and extreme debility.

The dose of Atropin is from ONE TWENTY-FOURTH to ONE-TWELFTH of ONE GRAIN. When the exhibition of this remedy produces dryness of the fauces, sparkling of the eyes, or dilation of the pupil, the dose must be diminished, or the remedy entirely laid aside for a time. Physicians will do well to triturate the Atropin with Asclepin, as the latter will in no case counteract the effects of the Atropin, but in view of its diaphoretic and neutralising properties, will materially enhance its action. It will also ensure a proper diffusion of the remedy, and enable the practitioner to more easily proportion and regulate the doses.

## CON. TINC. ATROPA BELLADONNA.

Derivation, properties and employment same as the Atropin. We give preference to this preparation of the Belladonna, as it is more diffusible, the dose is more easily proportioned than that of the powder, and is more readily prepared for local employment. Diluted with from ONE to EIGHT parts of water according to the extent and condition of the local affection, it is employed as an injection in painful neuralgic affections of the uterns and rectum, and as a local sedative over the seat of neuralgic pains, either by means of cloths saturated with the solution, or added to fomentations. When employed for injections, not more than TWICE or THRICE the quantity exhibited to the same patient at a dose should be administered.

The dose of the Tinc. is from ONE to FIVE DROPS.

# INDEX.

| | |
|---|---|
| Acids, Vegetable | 38 |
| Alkaloids | 40 |
| Amylum | 45 |
| Apotheme | 63 |
| Amygdalin | 68 |
| Asclepin | 122 |
| Ampelopsin | 153 |
| Alnuin | 266 |
| Apocynin | 349 |
| Aconitin | 433 |
| Atropin | 441 |
| Baptisin | 219 |
| Barosmin | 352 |
| Crude organic remedies | 17 |
| Constituents of plants | 31 |
| Cellulose | 42 |
| Cuticular or cork substance | 43 |
| Camphors | 55 |
| Caoutchouc | 58 |
| Coloring matters | 58 |
| Concentrated medicines | 70 |
| Concentrated medicines proper | 83 |
| Chemical transformations | 91 |
| Concentrated tinctures | 92 |
| Con. Tinc. Apocynum | 354 |
| " " Aconitum Napellus | 436 |
| " " Atropa Belladonna | 444 |
| " " Collinsonia | 412 |
| " " Cannabis Indica | 429 |
| " " Digitalis | 213 |
| " " Euonymus | 260 |
| " " Eupatorium Purpu. | 333 |
| " " Gelseminum | 139 |
| " " Gossypium | 432 |
| " " Hyoscyamus | 297 |
| " " Rhus Glabum | 218 |
| " " Scutellaria | 347 |
| " " Smilax | 402 |
| " " Strychnos Nux Vomica | 428 |
| " " Senecio | 120 |
| " " Veratrum | 325 |
| " " Xanthoxylum | 388 |
| Combinations | 95 |
| Cypripedin | 169 |
| Chimaphilin | 172 |
| Cornin | 272 |
| Caulophyllin | 279 |
| Corydalin | 334 |
| Chelonin | 181 |
| Collinsonin | 409 |
| Capsicum, oil of | 365 |

| | |
|---|---|
| Cerasein | 408 |
| Con. Comp. Stillingia Alterative | 424 |
| Colocynthin | 437 |
| Dextrine | 47 |
| Decoctions | 73 |
| Dioscorein | 176 |
| Digitalin | 201 |
| Daturin | 431 |
| Extractive substances | 59 |
| Emulsin | 68 |
| Extracts | 74 |
| " aqueous | 74 |
| " alcoholic | 76 |
| " hydro-alcoholic | 76 |
| " inspissated | 77 |
| " fluid | 77 |
| Euonymin | 257 |
| Eupatorin Perfo | 329 |
| Eupatorin Purpu. | 330 |
| Euphorbin | 371 |
| Erigeron, oil of | 261 |
| Erythroxylin | 431 |
| Fraserin | 378 |
| Fixed oils | 51 |
| Fluid extracts | 77 |
| Gum | 48 |
| Gum resins | 56 |
| Gelsemin | 129 |
| Geranin | 159 |
| Humus | 61 |
| Helonin | 185 |
| Hyoscyamin | 292 |
| Hydrastin | 359 |
| Hamamelin | 369 |
| Inulin | 47 |
| Infusions | 70 |
| Isolated preparations | 101 |
| Irisin | 355 |
| Jalapin | 284 |
| Juglandin | 337 |
| Leptandrin | 192 |
| Lupulin | 306 |
| Lycopin | 375 |

# INDEX.

Mucilage............................................ 49
Muci-resins......................................... 50
Macrotin............................................143
Myricin..............................................252
Menispermin........................................394

Neutral principles.......................... 41
Neutrals............................................. 59

Oils, fixed........................................51
 " volatile..................................... 53
Oleo-resins........................................ 56
Officinal preparations..................... 70
Oil of Erigeron..................................261
 " " Capsicum.................................365
 " " Stillingia................................302
 " " Populus...................................348
 " " Solidago..................................397
 " " Xanthoxylum.........................382
 " " Lobelia...................................422
Oleo. Resin of Lobelia......................414
Protein............................................. 44
Pectin............................................. 48
Populin............................................164
Phytolacin........................................287
Podophyllin......................................225
Prunin..............................................390

Resins............................................. 55
Resinoids......................................... 56

Rhusin..............................................214
Rumin...............................................276
Rhein...............................................489

Sugars............................................. 47
Syrups............................................. 79
Senecin............................................111
Stillingin........................................298
Scutellarin......................................344
Smilacin..........................................399
Sanguinarin......................................384
Strychnin.........................................427

Tinctures......................................... 79
Trillin............................................341

Vegetable bases................................ 40
Viscin............................................. 50
Volatile Oils.................................... 53
Viburnin..........................................269
Veratrin..........................................310

Wax................................................. 53
Wood substance................................. 43
Wine Tinc. Lobelia............................413

Xylogen........................................... 46
Xanthoxylin.....................................360
Xanthoxylum, oil of..........................382

www.ingramcontent.com/pod-product-compliance
Lightning Source LLC
Chambersburg PA
CBHW032130010526
44111CB00034B/576